SOLIDIFICATION

Edited by **Alicia Esther Ares**

Solidification
http://dx.doi.org/10.5772/intechopen.68508
Edited by Alicia Esther Ares

Contributors

Xiaolong Xu, Fawzy Samuel, Mohamed Ibrahim, Mohamed Abdelaziz, Bassam El-Eswed, Yandong Li, Huamei Duan, Cheng Peng, John Kenney, Juwon Lee, Socorro Valdez-Rodríguez, Hongfei Zheng, Xuesen Zeng, Hongjie Sun, Jianbao Zhang

Published by InTech
Janeza Trdine 9, 51000 Rijeka, Croatia

Publishing Process Manager Dajana Pemac
Technical Editor SPi Global
Cover InTech Design team

First published March, 2018
Printed in Croatia
Legal deposit, Croatia: National and University Library in Zagreb

Additional hard copies can be obtained from orders@intechopen.com

Solidification, Edited by Alicia Esther Ares
p. cm.
Print ISBN 978-953-51-3905-8
Online ISBN 978-953-51-3906-5

For EU product safety concerns:
IN TECH d.o.o., Prolaz Marije Krucifikse Kozulić 3, 51000 Rijeka, Croatia,
info@intechopen.com or visit our website at intechopen.com.

Meet the editor

Alicia Esther Ares is a headline professor of Materials Science at the Chemical Engineering Department, School of Sciences (FCEQyN), National University of Misiones (UNaM), Posadas, Misiones, Argentina, since December 2013. Also, she is an independent researcher at National Scientific and Technical Research Council (CONICET), Argentina, since January 2015. She was previously a research associate at CONICET (2008–2014) and an associate professor at UNaM (2007–2013) and has also been an assistant professor at UNaM (1989–2007). She graduated at the University of Misiones in 1992 and completed a PhD degree in Materials Science at Institute of Technology "Jorge Sabato," UNSAM-CNEA, Buenos Aires, Argentina. Later, she made a postdoctoral stay at the following institutions: Faculdade de Engenharía Mecânica; Departamento de Engenharía de Materiais; Universidade Estadual de Campinas, Campinas, São Paulo, Brazil (2001 and 2005–2006); Department of Materials Science and Engineering, University of Florida, Gainesville, Florida, United States (2002–20003); and Faculty of Sciences, National University of Misiones, Posadas-Misiones, Argentina (2003–2004).

She has a 29-year teaching experience both at the undergraduate and at the graduate levels.

Her research interests lie in the following areas:

- Solidification thermal parameters, mechanical properties, and corrosion resistance of different alloys and composite materials
- Solidification structures and properties of alloys for hard tissue replacement
- Metallic material selection for the management of biofuels
- Synthesis and characterization of nanostructured coatings, membranes, and templates of aluminum and zinc oxides
- Fabrication and characterization of nanostructured titanium and iron oxide coatings for water treatment systems based on advanced oxidative and reductive processes
- Natural products as corrosion inhibitors of metallic materials

Her articles are published in well-established international and Argentinian journals.

Contents

Preface

In a solidification process, the material is first heated to melt and then is poured into a cavity of the mold. When the molten metal is in the mold, it begins to cool. When the temperature drops below the freezing point (melting point) of the material, solidification starts.

Solidification involves a change of phase of the material and differs depending on whether the material is a pure element or an alloy. A pure metal solidifies at a constant temperature, which is its melting point (freezing point). For alloys, the solidification occurs over a temperature range depending upon the composition.

Similar to the processing of metals and alloys, the thermoplastic processing depends critically on the ability to melt and process them via different processes. Many polymers do not crystallize but solidify when cooled.

Inorganic glasses do not crystallize simply. The melt is ever so often excessively viscous, and the diffusion is too slow for crystallization to advance during solidification.

The book *Solidification* is logically developed through a careful presentation of the relevant theories and models of solidification occurring in a variety of materials. Mathematicians, chemists, physicists, and engineers concerned with melting/freezing phenomena will also find this book to be valuable.

Section I: Effect of Microalloying Elements on the Heat Treatment Response and Tensile Properties of Al-Si-Mg Alloys

Chapter 1 (M.F. Ibrahim, M.H. Abdelaziz, H.W. Doty, S. Valtierra, and F.H. Samuel) investigates the effects of alloying elements and heat treatment conditions on the microstructure and mechanical behavior of nonmodified and Sr-modified 356- and 357-type alloys by examining several factors.

Section II: Rapid Solidification of Undercooled Melts

In Chapter 2 (Xiaolong Xu, Hua Hou, and Feng Liu), the authors experimentally investigate the microstructural evolution of the Ni-20at.%Cu alloys as a function of initial undercooling and the physical mechanisms of the grain refinements occurring at low undercooling regimes. In combination with the current dendrite growth model, they theoretically analyze the dendrite remelting in the undercooled alloys by an extended chemical superheating model for nonequilibrium solidification of undercooled binary single-phase alloys.

Section III: Coupled Model of Precipitates and Microsegregation during Solidification

In order to understand precipitate behavior during the solidification process of microalloyed steel containing rare-earth elements, thermodynamic calculation and equilibrium experiments were adopted to simulate the solidification process. The experiments and analysis in Chapter 3 (Yandong Li, Huamei Duan, and Cheng Peng) include FactSage calculations, equilibrium experiments, thermodynamic model establishment, precipitate prediction, and dynamic experiments.

Section IV: Physics of Solidification and Microstructure: A Study of AlMg and AlMgSi by Vortex Method

Chapter 4 (S. Valdez Rodríguez, L. Martínez-Gómez, and M.I-Pech Canul). The microstructure characteristic is essential for the properties of metallic materials, including binary, ternary, or eutectic alloys.

Consequently even with significant progress on microstructural evolution, numerous challenges still exist for making the internal structure to act synergistically with matrix alloy chemical components.

In this chapter, the authors review the correlation between equilibrium and nonequilibrium effects, surface energy, and chemical potential as the time of the structure formation is probably the key to solidification that can help to predict the complex of microstructures.

For the case of multiphase solidification, AlMgSi alloy involves thousands of atoms, at the atomic scale, that transit to microstructures, while the solid-liquid phase transformation occurs. The analysis outlines the role of physic solidification and the atomic arrangement influencing mechanical hardness properties and degradation resistance.

Section V: Solidification versus Adsorption for Immobilization of Pollutants in Geopolymeric Materials

In Chapter 5 (B. I. El-Eswed), the author reviews the latest development in the fields of immobilization of heavy metals, radio-actinides, dyes, and other hazardous wastes in geopolymeric materials. Two techniques are investigated and compared: the first is solidification and the second is adsorption. Geopolymer (GP), as host for solidification and as adsorbent for adsorption, is a class of amorphous three-dimensional aluminosilicate binder material, produced by the reaction of an aluminosilicate source (metakaolin or fly ash) with a highly concentrated aqueous alkali metal silicate or hydroxide. The chapter reviews and evaluates solidification and adsorption of different kinds of pollutants in GP matrix. The two techniques are compared regarding efficiency or capacity of GP, reversibility of process, mechanism of adsorption or solidification, leaching of contaminants, stability of the material, cost, energy, and environmental impact.

Section VI: The Application of Carbon Fiber Composites in Cryogenic Fuel Tanks

The applications of carbon fiber composites in liquid hydrogen (LH2) and liquid oxygen (LOX) fuel tanks are introduced in Chapter 6 (H. Zheng, X. Zeng, J. Zhang, and H. Sun). The materials, processing, and design of DC-XA LH2 tank, X-33 LH2 tank, SLI LH2 tank, and CCTD program tank are discussed. Lockheed Martin LOX tank and Space X Lox tank are introduced by the authors. Technology development, materials development, and development trend of cryogenic fuel tanks are discussed. Thin-ply hybrid laminates and out-of-autoclave tanks are projected for future space missions.

Section VII: Clathrate Hydrates

Chapter 7 (J. Lee and J. W. Kenney, III) In this chapter, a brief historical review of the formation, structure, and uses of clathrate hydrates forms the backdrop for a discussion on modern scientific investigations of these solids employing spectroscopy, structure determination methods, isotopic studies, computational-theoretical modeling, and interrogations of guest-host interactions via special guests. For example, the use of colored halogens in clathrate hydrate hosts enables UV-visible spectroscopic methods to be employed to study the clathrate hydrate structure.

Prof. Dr. Alicia Esther Ares
Materials Institute of Misiones (CONICET–UNaM)
Posadas, Misiones, Argentina

Chapter 1

Effect of Microalloying Elements on the Heat Treatment Response and Tensile Properties of Al-Si-Mg Alloys

Mohamed F. Ibrahim, Mohamed H. Abdelaziz,
Herbert W. Doty, Salvador Valtierra and
Fawzy H. Samuel

Additional information is available at the end of the chapter

http://dx.doi.org/10.5772/intechopen.70665

Abstract

This study was carried out on a series of heat-treatable Al-Si-Mg alloys to determine the effects of Fe, Mg, Sr and Be addition on their microstructural characteristics and tensile properties. The results showed that the eutectic temperature was reduced by 10°C with 0.8 wt% Mg addition. The solidification curves and first derivatives of Sr-free alloys with high Fe and Mg contents revealed a peak at 611°C consequent to the formation of a script-like Be-Fe (Al8Fe2BeSi) phase, which was very close to the peak for α-Al. The morphology of the β-iron platelets underwent changes due to their dissolution, thinning, necking, and fragmentation with increase in solutionizing time. Increased Mg contents are beneficial to the tensile properties unlike the detrimental effect of increasing Fe contents. Additions of Be and Sr noticeably improved the properties at the same Fe and/or Mg contents, the enhancements being markedly observed at higher Mg contents and reduced Fe levels. At high Fe levels, addition of Be is preferable as it neutralizes the deleterious effects of Fe in these alloys; however, addition of 500 ppm Be is inadequate for interacting with other alloying elements.

Keywords: aluminum alloys, Minitab, scattered plot, intermetallics, Si coarsening

1. Introduction

The main objectives of heat-treating cast Al-Si-Mg alloys include homogenization, stress relief, improved dimensional stability, and optimization of the strength and ductility parameters. The T6 heat-treated Al-Si-Mg alloys have an optimum combination of strength and ductility.

The typical heat treatment specification of a T6 temper consists of solid solution treatment and quenching, followed by aging. The main effects of solution heat treatment are the dissolution of Mg_2Si particles, the homogenization of the casting, and modification of the morphology of eutectic silicon through fragmentation and spheroidization at critical temperatures. The recommended solution temperature for 356 and 357 alloys is 540 ± 5°C [1, 2].

The purpose of quenching is to preserve the solid solution formed at the solution heat treatment temperature by rapid cooling to room temperature. The quenching medium and rate are the parameters which control the mechanical properties. The highest strength can be ensured when the material is subjected to a rapid quenching rate. Aging is the final stage of heat treatment in cast Al-Si-Mg alloys. The aging can be natural or artificial. Alloy mechanical properties depend on both aging temperature and time. The main objective of artificial aging is to heat the as-quenched castings to an intermediate temperature between 150 and 200°C for 4–8 h in order to precipitate the excess solutes which were supersaturated in α-Al during the solution heat treatment process. The improvement achieved in the mechanical properties during artificial aging is due to the precipitation of metastable phases from the supersaturated solution. Al-Si-Mg alloys fulfill the following precipitation sequence [3]:

SS → needle-shaped GP zones → rod-like β' precipitates → platelets of Mg_2Si.

For Al-Si-Mg alloys, the solubility of Mg and Si in the Al matrix decreases with temperature. In order to obtain a maximum concentration of Mg and Si particles in solid solution, the solution temperature should be close to the eutectic temperature. For the 356 and 357 alloys, the solution temperature is 540 ± 5°C. At this temperature, about 0.6% Mg can be placed in solid solution. The dissolution of Mg_2Si into Mg and Si occurs in the two alloys at 475 and 540°C, respectively [4].

An investigation [5] was carried out on the effects of T6 and T4 tempers on the tensile properties of nonmodified and Sr-modified A356 alloys obtained from permanent mold and sand mold castings. The results showed that the yield strength is not appreciably influenced by the change in Si-particle characteristics. The ultimate tensile strength (UTS) and %El increases significantly with increase in the solution treatment time at 540°C. Also it was observed that upon modification with Sr the fracture mode changes from brittle to ductile especially in sand castings. Both non-modified and Sr-modified alloys obtained from permanent mold casting, however, showed ductile fracture mode. All the improvements in the tensile properties reported upon were related mainly to the changes which occurred in the Si particle aspect ratio and particle size during solution treatment.

Yoshida and Arrowood [6] studied the effects of varying the solutionizing parameters (time and temperature) of a T6 traditional treatment on the mechanical properties (i.e., hardness, ductility, and ultimate tensile strength) of Sr-modified and nonmodified permanent mold cast A356 alloys. The investigated solution treatment times were 2, 4, 8, 16, and 32 h, while the solution treatment temperatures were 520 and 540°C, where the aging treatment was kept unchanged at 160°C for 6.5 h. The highest hardness was obtained at the shortest solutionizing time of 2 h for both the unmodified and modified A356 alloy, while the highest ductility was not reached until the samples were solutionized for 8 h at the same temperature. A slight change in solutionizing temperature did not cause much variation in hardness, ductility, or UTS. It may also be concluded that the Sr-modified samples exhibit higher ductility than the unmodified ones under all the heat treatment conditions reported in this study.

A valuable study was carried out by Moustafa et al. [7, 8] on A413. 1 Al-Si eutectic alloys, where they reported that alloys with Mg contents suffered reduction in hardness, yield strength, and ultimate tensile strength values following the addition of Sr. They explained this observation in terms of retarded precipitation of Mg_2Si particles during the aging treatment owing to longer incubation period preceding the commencement of precipitation, irrespective to the solutionizing time [8, 9]. Also, alloying element additions of Mg and Be, resulted in improving the hardness and strength of the base alloy, especially in the T6 condition, where adding small amount of Be (~0.02%) prevented Mg oxidation (i.e., formation of MgO and $MgAl_2O_4$ (spinel)) during melting, so that the hardness increased slightly.

The main purpose of this study, on Al-Si-Mg type 356 and 357 casting alloys, is to investigate the effects of alloying elements and heat treatment conditions on the microstructure and mechanical behavior of nonmodified and Sr-modified 356- and 357-type alloys by examining the following factors:

1. The influence of alloying elements on the aging behavior of alloy castings investigated in relation to:
   a. Fe content, Sr-modification, grain refining, and addition of Mg and Be; and
   b. Solution heat treatment and aging parameters.
2. Microstructural analysis of the precipitated phases during the solidification process.
3. Correlating the results obtained from the tensile testing with the microstructural analysis to determine the effects of each alloying element, intermetallic phase, changes in the morphology, solutionizing parameter, and aging condition on the mechanical properties of the alloys investigated.

2. Experimental procedure

The chemical composition of the B356.2 alloy used in the present study is shown in **Table 1**. The Mg level of the alloy was increased by adding pure Mg to the alloy melts to obtain Mg levels of 0.4, 0.6, and 0.8 wt%. The Fe and Be were added in the form of Al-25% Fe and Al-5% Be master alloys, respectively, to the alloy melt to obtain Fe levels of 0.09, 0.2, and 0.6 wt% and a Be level of 0.05 wt%. The Sr and Ti were added in the form of Al-10% Sr and Al-5% Ti master alloys, for Sr-modification and grain refining purposes, respectively, to the alloy melts to obtain levels of 0.02 wt% Sr and 0.15 wt% Ti.

AA alloy no.*	Chemical composition (wt%)**					
	Si	Fe	Cu	Mg	Zn	Al
B356.2	7.0	<0.06	<0.025	0.35	<0.001	Bal

*AA, aluminum association.
**Unlisted aluminum or impurities.

Table 1. Chemical composition of the B356.2 alloy (wt%) [8].

An electrical resistance furnace with a 70-kg crucible capacity was used to prepare 60 kg of each alloy. The melt was kept at 750 ± 5°C as the melting temperature at which additions of magnesium, iron, beryllium, strontium, and titanium were made. These measured additions were added to the melt using a perforated graphite bell. Before casting, the melt was degassed using pure and dry argon for 20 min using a graphite impeller in order to clean the melt from hydrogen and inclusions. For the purposes of thermal analysis, one sample of each alloy composition was prepared using the arrangement shown in **Figure 1**; moreover, two samplings for chemical analysis were also taken from each melt, one before the start of casting and one at the end of casting.

The chemical analysis was carried out using arc spark spectroscopy at GM facilities in Milford, NH. The actual chemical composition of each of the alloys prepared is shown in **Table 2**, representing average values taken over three spark measurements made for each chemical analysis sample. The alloys A1 through C3B shown in the table represent 18 non-modified alloys

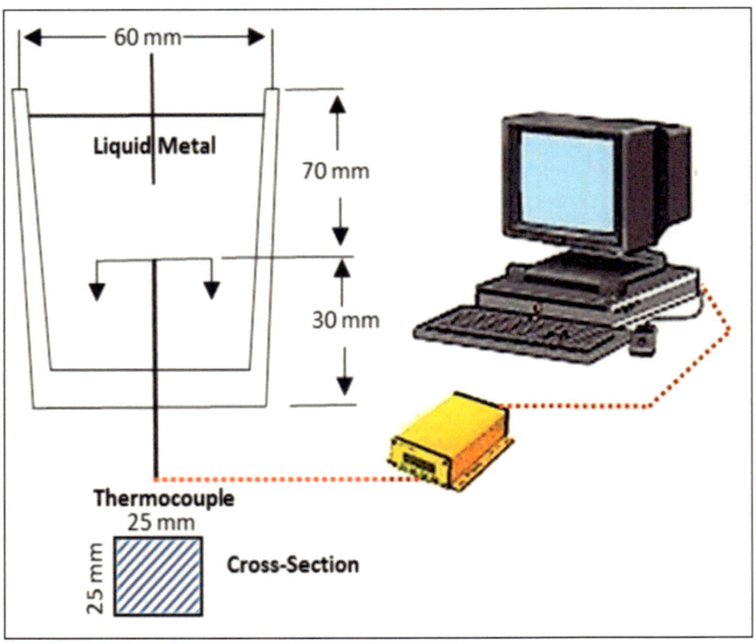

Figure 1. Schematic drawing showing the graphite mold used for thermal analysis.

Alloy code	Element concentration (wt%)								
	Si	Fe	Mg	Cu	Mn	Ti	Sr	Be	Al
A1*	7.02	0.101	0.376	0.050	0.010	0.143	0.001	–	Bal.
B1	7.00	0.203	0.383	0.008	<0.001	0.151	0.001	–	Bal.
C1	6.97	0.630	0.372	0.011	0.002	0.174	0.001	–	Bal.
A2	6.99	0.113	0.590	0.042	0.008	0.164	0.001	–	Bal.
B2	7.03	0.210	0.570	0.009	0.002	0.179	0.001	–	Bal.
C2	6.88	0620	0.540	0.010	0.002	0.172	0.001	–	Bal.
A3	7.08	0.123	0.760	0.034	0.007	0.192	0.002	–	Bal.
B3	7.35	0.226	0.750	0.009	0.002	0.197	0.001	–	Bal.
C3	7.07	0.630	0.750	0.010	0.003	0.183	0.002	–	Bal.
A1B*	7.33	0.093	0.368	0.052	0.006	0.180	0.001	0.034	Bal.
B1B	6.61	0.197	0.474	0.011	<0.001	0.180	0.001	>0.036	Bal.
C1B	6.89	0.670	0.510	0.011	0.002	0.186	0.003	>0.036	Bal.
A2B	7.26	0.081	0.530	0.006	<0.001	0.173	0.001	>0.036	Bal.
B2B	6.60	0.184	0.710	0.007	<0.001	0.177	0.001	>0.036	Bal.
C2B	6.22	0.680	0.700	0.007	0.002	0.183	0.001	>0.036	Bal.
A3B	6.50	0.127	0.980	0.034	0.006	0.203	0.001	0.026	Bal.
B3B	6.38	0.196	0.960	0.009	0.001	0.210	0.001	>0.036	Bal.
C3B	5.96	0.560	0.920	0.009	0.003	0.221	0.002	>0.036	Bal.
A1S	7.06	0.103	0.338	0.046	0.010	0.139	0.016	–	Bal.
B1S	7.18	0.217	0.354	0.014	0.002	0.165	0.019	–	Bal.
C1S	7.18	0.660	0.349	0.007	0.002	0.172	0.017	–	Bal.
A2S	7.07	0.105	0.520	0.033	0.002	0.156	0.018	–	Bal.
B2S	7.26	0.217	0.530	0.009	0.001	0.169	0.015	–	Bal.
C2S	7.17	0.640	0.530	0.008	0.002	0.154	0.017	–	Bal.
A3S	7.98	0.101	0.820	0.035	0.002	0.152	0.013	–	Bal.
B3S	7.26	0.185	0.730	0.030	0.002	0.171	0.009	–	Bal.
C3S	6.84	0.710	0.860	0.034	0.005	0.169	0.006	–	Bal.
A1BS	6.54	0.122	0.640	0.041	0.007	0.227	0.048	>0.036	Bal.
B1BS	6.12	0.190	0.491	0.009	<0.001	0.217	0.027	>0.036	Bal.
C1BS	7.78	0.810	0.530	0.084	0.012	0.210	0.016	>0.036	Bal.
A2BS	7.64	0.123	0.780	0.010	0.002	0.243	0.024	>0.036	Bal.
B2BS	5.93	0.194	0.730	0.007	<0.001	0.227	0.023	>0.036	Bal.
C2BS	7.06	0.670	0.710	0.007	0.002	0.222	0.022	>0.036	Bal.

Alloy code	Element concentration (wt%)								
	Si	Fe	Mg	Cu	Mn	Ti	Sr	Be	Al
A3BS	7.26	0.105	0.950	0.020	0.002	0.188	0.023	>0.036	Bal.
B3BS	6.33	0.194	0.950	0.021	0.002	0.189	0.017	>0.036	Bal.
C3BS	6.21	0.690	0.850	0.033	0.005	0.207	0.014	>0.036	Bal.

Table 2. Average chemical composition (wt%) of the alloys studied.

while their Sr-modified counterparts were coded A1S to C3BS, respectively. The Sr level ranged from 0 to 0.02 wt% in these alloys. The codes A, B, C, 1, 2, etc. are explained below.

A = 0.1% Fe, B = 0.2% Fe, C = 0.6% Fe, 1 = 0.4%Mg, 2 = 0.6% Mg, 3 = 0.75% Mg, S = Sr,*B = Be.

The various alloys were used to prepare castings from which test bars were obtained for tensile testing purposes. With this aim in mind, the degassed molten metal was carefully poured into an ASTM B-108 permanent mold preheated to 450°C, to obtain castings for tensile testing. **Figure 2a** shows the actual casting obtained whereas **Figure 2b** shows the dimensions of the tensile bars.

Following casting in the ASTM B-108 permanent mold, the tensile bars were divided into 13 different bundles (5 bars/bundle); each bundle was subjected to certain heat treatment conditions as follows: 1 bundle was kept in the as-cast condition; 1 bundle was solution treated at 540°C for 5 h, then quenched in warm water at 65°C, and was kept in the as-solutionized condition; another bundle was kept also in the as-solutionized condition, however, after solutionizing for 12 h at 540°C followed by quenching in warm water at 65°C; 5 bundles were solution heat-treated at 540°C for 5 h, then quenched in warm water at 65°C followed by artificial aging at 160°C for 2, 4, 6, 8, and 12 h, respectively; the remaining 5 sets (bundles) were solution heat-treated at 540°C for 12 h, then quenched in warm water at 65°C followed by artificial aging at 160°C for 2, 4, 6, 8, and 12 h, respectively. The solution and aging heat-treatments were carried out in a forced-air Blue M electric furnace equipped with a programmable temperature controller, accurate to ±2°C. The aging delay was less than 10 s. For each individual heat treatment, five test bars were used.

For each condition, the test bars were pulled to fracture at room temperature at a strain rate of 4×10^{-4}/s using an MTS Servohydraulic mechanical testing machine. During the course of the tensile test, an attachable strain-gauge extensometer was connected to the gauge section of the tensile bars to measure the percentage elongation. The reported tensile data were the percentage elongation to fracture (%EF), 0.2% offset yield strength, and ultimate tensile strength; these data were reported as the average values of five data sets obtained from pulling five bars per condition.

Samples for microstructural analysis were taken from both the tensile-tested bars ~10 mm below the fracture surface and the as-cast thermal analysis castings, which were sectioned to study each alloy condition, that is, for the tensile-tested bars, one sample was used in the as-cast condition, while the other two were solution heat-treated (540°C/5 h and 540°C/12 h) beside the as-cast thermal analysis sample. The microstructures of the polished sample

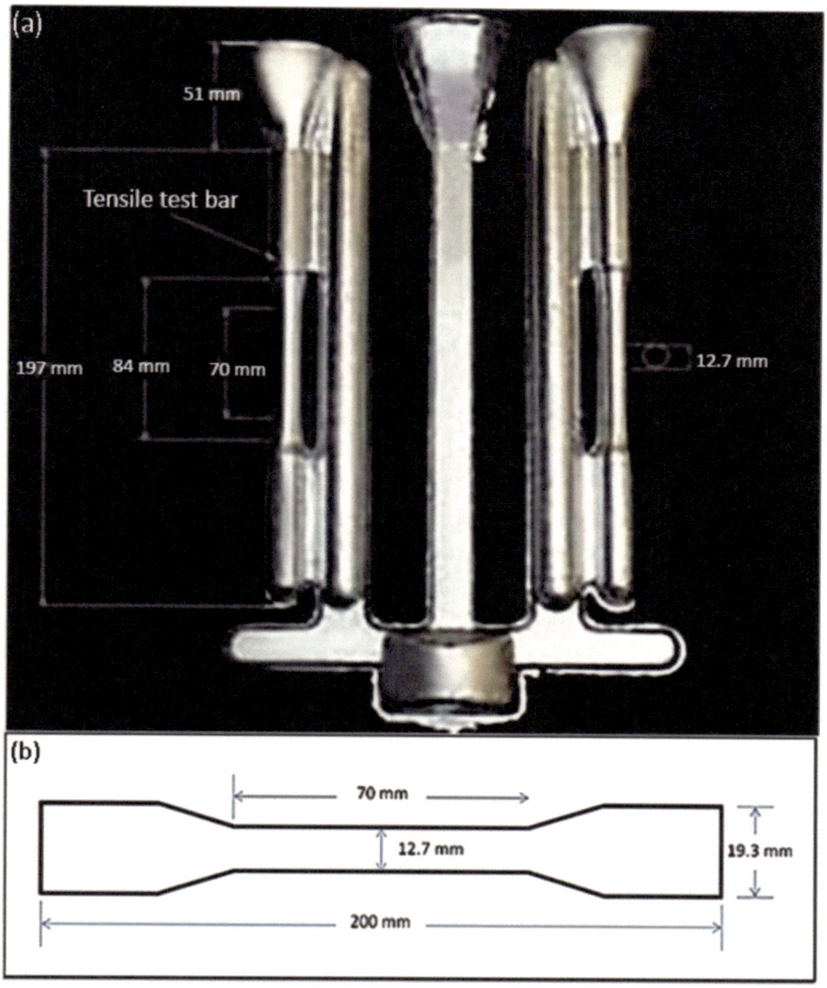

Figure 2. (a) ASTM B-108 permanent mold used for casting tensile test bars and (b) dimensions of the tensile test bar (in mm).

surfaces were examined using an optical microscope linked to a Clemex image analysis system, and a Hitachi S-4700 field emission scanning electron microscope (FE-SEM), equipped with a standard secondary electron detector (SE), a backscatter electron detector (BSD), and an energy dispersive X-ray spectrometer (EDS).

3. Results and discussion

3.1. Eutectic Si particles

Measurements of the dendrite arm spacing (DAS) showed that the DAS was about 24 μm in the as-cast tensile bars. The Si particle measurements for the A1 base alloy, containing low levels of Mg and low Fe, and modified base alloy A1S, show that the Si particle area decreased from 28.4 to 0.67 μm², the Si particle length decreased from 10.8 to 1.25 μm (as can be seen from **Figure 3**), the aspect ratio decreased from 3.78 to 1.84, while the roundness ratio increased from 18.9 to 50.3% upon addition of Sr to the A1 base alloy in the as-cast condition [9–14].

The addition of Be in alloy A1B1 resulted in reduced silicon particle area of 10.3 μm², a reduced silicon particle length of 6.6 μm, a lower value of aspect ratio (2.73), and an increased roundness ratio up to a value of 35.7%. The mutual addition of Sr and Be (alloy A1BS) led to a reduction in silicon particle area (1.31 μm²), silicon particle length (1.84 μm), and the aspect ratio (1.94); however the roundness ration increased to 47.6%. By increasing the Fe and Mg content (alloy C3), the silicon particle area decreased to 6.9 μm², the silicon particle length and aspect ratio reduced to 6.44 μm and 3.01, respectively, while the roundness ratio increased to 24.6%. The combined addition of strontium, beryllium, iron, and magnesium (alloy C3BS) reduced the silicon particle area to 1.53 μm², the silicon particle length to 2 μm, the aspect ratio to 1.99 and increased the value of the roundness ratio to 46.7%. The aforementioned values of the silicon particle characteristics highlight the modification effect of Sr and the partial modification effect of both Mg and Be. However, with increasing Fe levels it seems that most of the Be reacts with the Fe forming a Be-Fe phase (Al8Fe2BeSi). The results for the average Si particle length are summarized in **Figure 3**.

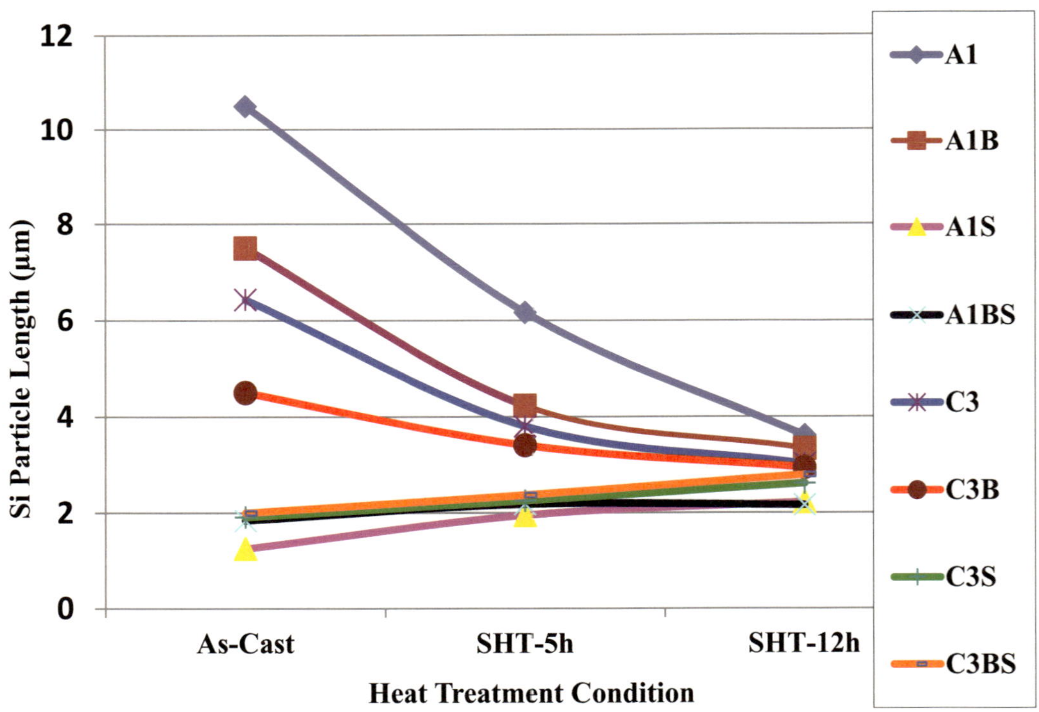

Figure 3. Average length of eutectic silicon particles with different solution heat treatment conditions.

Figure 4 shows the morphology of the eutectic Si particles in the A1 alloy, deeply etched in hydrofluoric acid (HF) solution. As can be seen in **Figure 4a**, the Si particles precipitated in the form of short platelets with sharp angles (white arrow). Solutionizing at 540°C resulted in fragmentation of the Si particles as demonstrated by the circled area in **Figure 4b**. As inferred from **Figure 4c**, there is a large difference in the particle sizes (see arrow) caused by dissolution of these particles in the aluminum matrix leading to coarsening of the other particles and hence the reported standard deviation (100%). Another observation that could be made from **Figure 4c** is that the new fragments of the Si platelets are still maintaining the platelet shape as shown by the more elongated particles in the figure. This process continued even after solutionizing for 400 h at 540°C as displayed in **Figure 4d–f**.

Figure 5 shows the microstructure of A1S alloy. As expected, addition of Sr 160 ppm resulted in changing the morphology of the Si particles to fibrous as displayed in **Figure 5a**. The microstructure of the A1S alloy following solutionizing at 540°C for 5 h is presented in **Figure 5b**. Three observations could be made from this figure:

i. Necking of the Si particles-solid arrow

ii. Dissolution of some particles in the matrix-broken arrow, in keeping with the: Ostwald ripening mechanism [15] (see **Figure 6**).

iii. Coarsening of other particles by collision.

Increasing the solutionizing time to 12 h at 540°C led to partial spheroidization as shown by the solid arrows in **Figure 5c** where the new particles are having multiple sides. The white area circled in **Figure 5c** points to particle collision. **Figure 5d** is an enlarged portion of **Figure 5c** showing the possibility of the fusion of some particles in the background to the left, leading to a coarser one. This process of coarsening was observed to continue up to 400 h at 540°C, **Figure 5d–f**. It should be noted here that the Si particles are not spherical; rather, they possess multiple faces (**Figure 5f**).

3.2. Precipitation of Mg_2Si

Samuel et al. [16] observed that any Sr-modified microstructure is clearly affected when magnesium is present. Microstructural parameters as obtained from image analysis, such as silicon particle size and aspect ratio, were found to increase with an increasing Mg content, subsequently becoming increasingly inhomogeneous. The reason for the deterioration in modification is believed to be the formation of intermetallic phases of the type $Mg_2SrAl_4Si_3$, where the addition of Mg also lowers the eutectic temperature - the eutectic temperature decreases with increasing Mg content. It was also reported, however, that a magnesium content of ~1 wt% itself acts as a refiner for the eutectic silicon in unmodified Al-Si alloys.

Ibrahim et al. [17] showed that the addition of up to 0.5 wt% of Mg to molten 319 type alloys results in the formation of Mg_2Si; the Mg-rich phase commonly exists as rounded black particles close to the eutectic silicon particles. It was also reported that the addition of Mg results in remarkable fragmentation/modification of the eutectic silicon particles as well as transformation of a large portion of the harmful needles of β-Al_5FeSi Fe phase into the less detrimental Chinese script-like phase with a composition close to $Al_8Mg_3FeSi_6$.

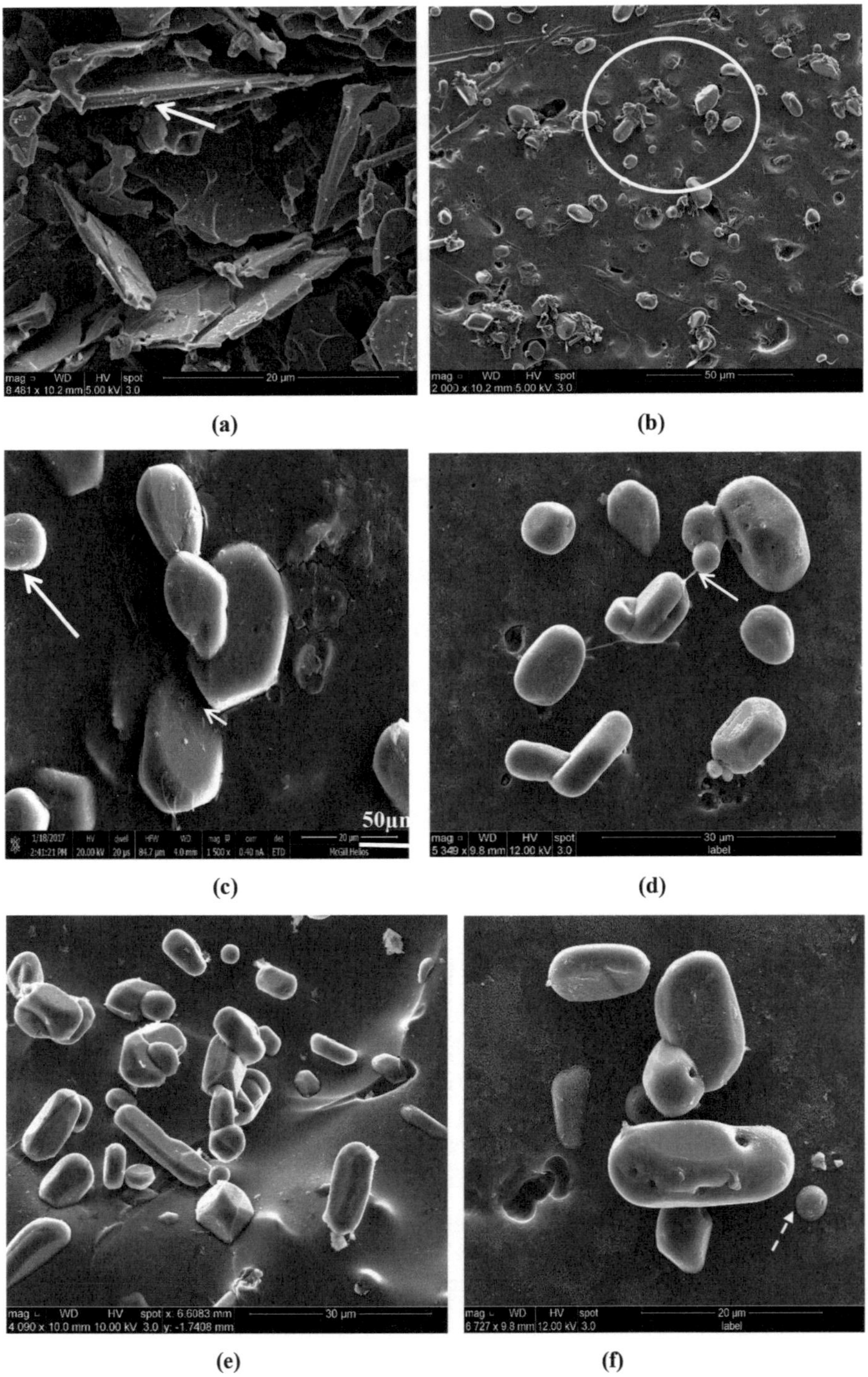

Figure 4. Variation in eutectic Si particles in A1 alloy as a function of heat treatment: (a) as cast, (b and c) after 5 h at 540°C, (d) after 12 h at 540°C, (e) after 200 h at 540°C, and (f) after 400 h at 540°C. The white arrow in (f) reveals the dissolution of fine particles even after the long solutionizing time at 540°C.

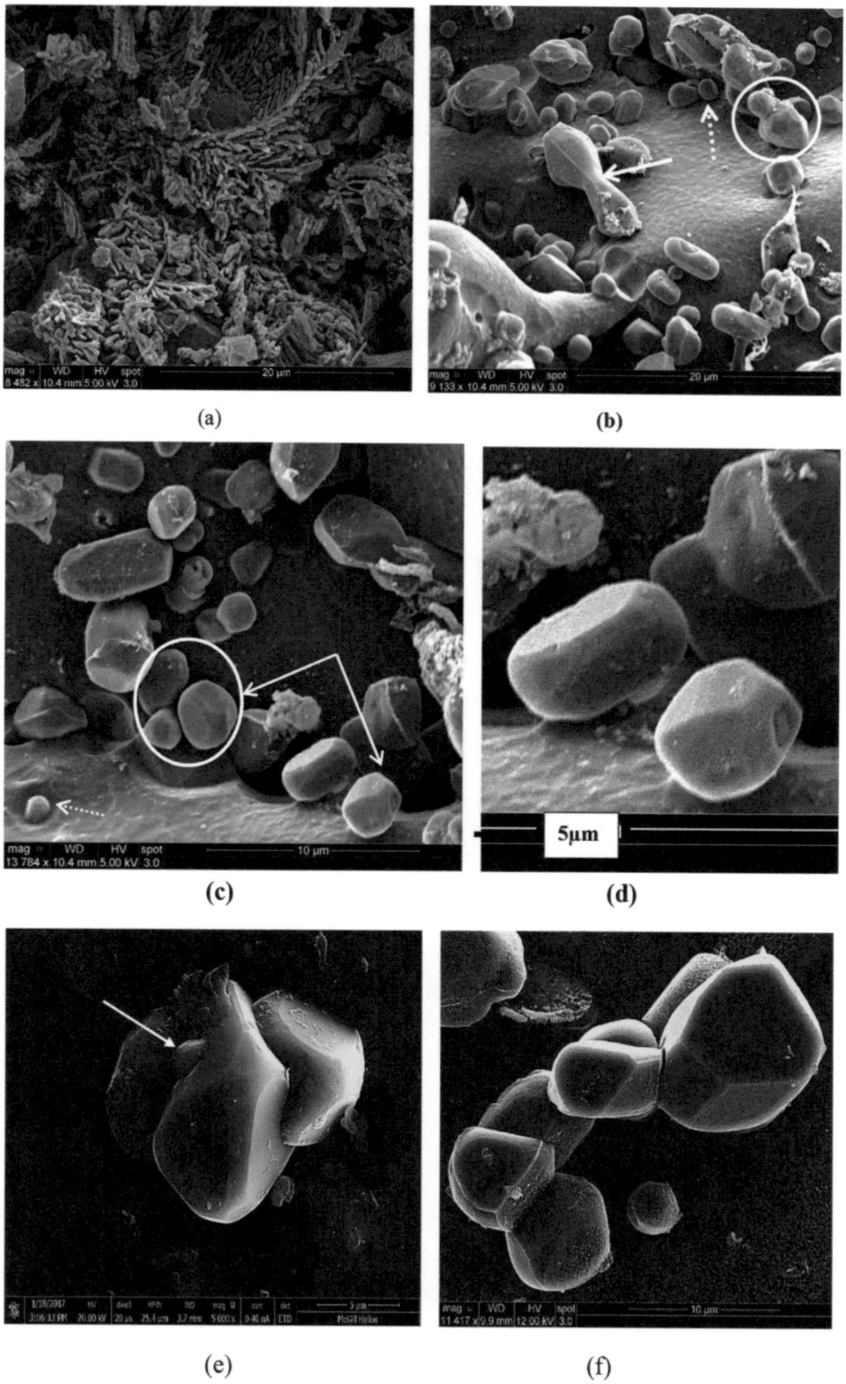

Figure 5. Variation in eutectic Si particles in A1S alloy as a function of heat treatment: (a) as cast, (b) after 5 h at 540°C, (c and d) after 12 h at 540°C, and (e) after 200 h at 540°C, and (f) after 400 h at 540°C broken arrow in (b) points to the presence of very small Si particles, whereas the solid arrow in (e) indicates the complete fusion of a Si particle in a cluster of particles.

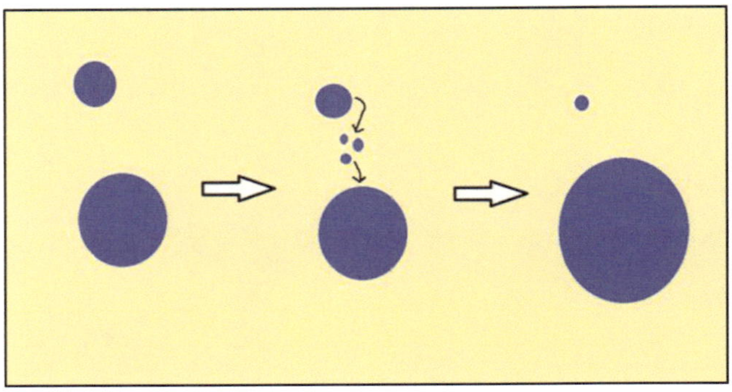

Figure 6. Schematic diagram showing coarsening of Si particles according to Ostwald ripening mechanism [15].

The heat treatment for Al-Si-Mg alloy is T6 which consists of solution heat treatment and natural or artificial aging. A solution treatment between 400 and 560°C, results in the dissolution of the hardening phase (Mg_2Si) in the aluminum matrix. However, the solutionizing temperature is limited by the eutectic phase melting temperature. The alloy thereafter is aged at low temperature (150–200°C) for precipitation of the hardening compounds which improves the mechanical properties of the aluminum matrix [18]. The primary purpose of the long solution treatment is to thermally alter silicon particle characteristics [19]. During T6 treatment, the Mg and Si which are in solid solution precipitate as Mg_2Si during the aging treatment [13].

Figure 7a reveals the precipitation of Mg_2Si in the form of Chinese script (black) following the formation of π-$Al_8Mg_3FeSi_6$ in 356 alloy solidified at 0.8°C/s (A1 alloy). The identity of these two phases was confirmed from their corresponding EDS spectra shown in **Figure 7b** and **c**, respectively. The microstructure of the A1 alloy following solutionizing at 540°C for 12 h is shown in **Figure 8a**, revealing the presence of scattered particles. These particles have been identified as Si particles [20]. Following aging at 160°C for 12 h resulted in the dense precipitation of Mg_2Si particles in the form of short rods-**Figure 8b**. Obviously, the density of the Mg_2Si precipitation is controlled by the amount of added Mg and the presence of Sr [21]. Another parameter to be considered during long solutionizing times is the decomposition of the π-phase into the β-phase as demonstrated in **Figure 9** [22].

3.2.1. Tensile properties

Tensile properties of reference alloys A1, A3, and C3 in the as-cast and different heat treatment conditions applied are summarized in **Table 3**. In the as-cast samples, increasing Mg content from 0.4 wt% (base alloy A1) to 0.8 wt% (alloy A3) resulted in slightly increasing the strength due to the partial modification effect of Mg as well as the transformation of β-phase to π-phase, that is, most of the Fe-intermetallic phases precipitated in the form of the π-phase in spite of the low Fe content (0.09 wt%), whereas increasing the Fe level up to 0.6 wt% in the high Mg-containing alloy A3 i.e. alloy C3 resulted in decreasing the UTS somewhat, because of the growing formation of Fe-intermetallic phases. For the three reference alloys, percent elongation was decreased.

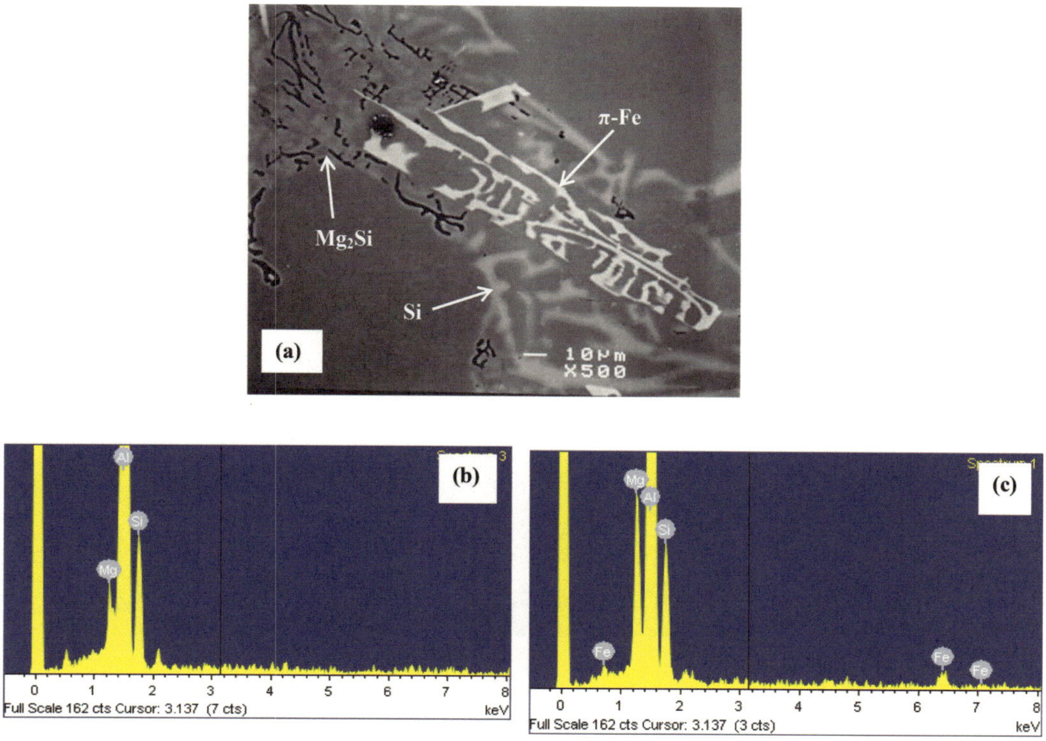

Figure 7. Precipitation of Mg_2Si and π-Fe in 356 alloy solidified at 0.8°C/s: (a) backscattered electron image, (b) EDS spectrum corresponding to Mg_2Si phase, and (c) EDS spectrum corresponding to π-phase.

Applying 5 h solution heat treatment at 540°C appeared in a noticeable increase in the ultimate strength (UTS) values, (59 MPa) for alloys A1 and A3, and (39 MPa) for alloy C3, due to the fragmentation and spheroidization of Si particles and spheroidization of undissolved π-phases as well as dissolving of Mg_2Si in the matrix and the decomposition of the π-phase to β-phase which further fragmented during solution heat treatment, thereby improving the tensile properties. For both solutionizing times, the elongation values obtained were two times or more higher than those observed in the as-cast condition for alloys A1, A3, and C3, indicating improved ductility of the solution heat-treated samples. Similar results regarding the addition of alloying elements and solution heat treatment parameters have been reported in a number of studies [23–30].

The tensile data of the 31 alloys used in the present work was classified into 4 series:

1. Fe-Mg series.

2. Fe-Mg-Be series.

3. Fe-Mg-Be-Sr series.

4. Fe-Mg-Sr series.

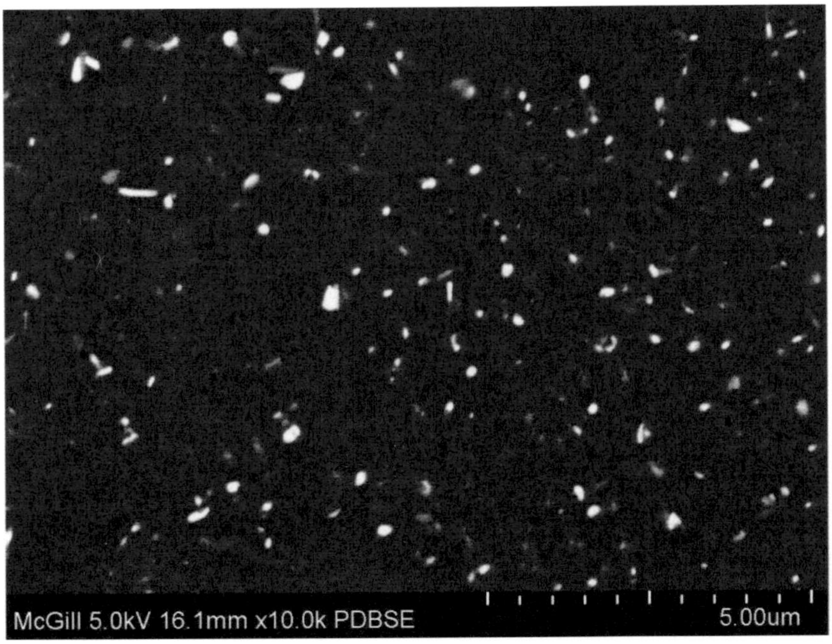

(a)

(b)

Figure 8. Backscattered electron images of the precipitation observed in A1 alloy tensile bars: (a) T4 condition and (b) T6 condition.

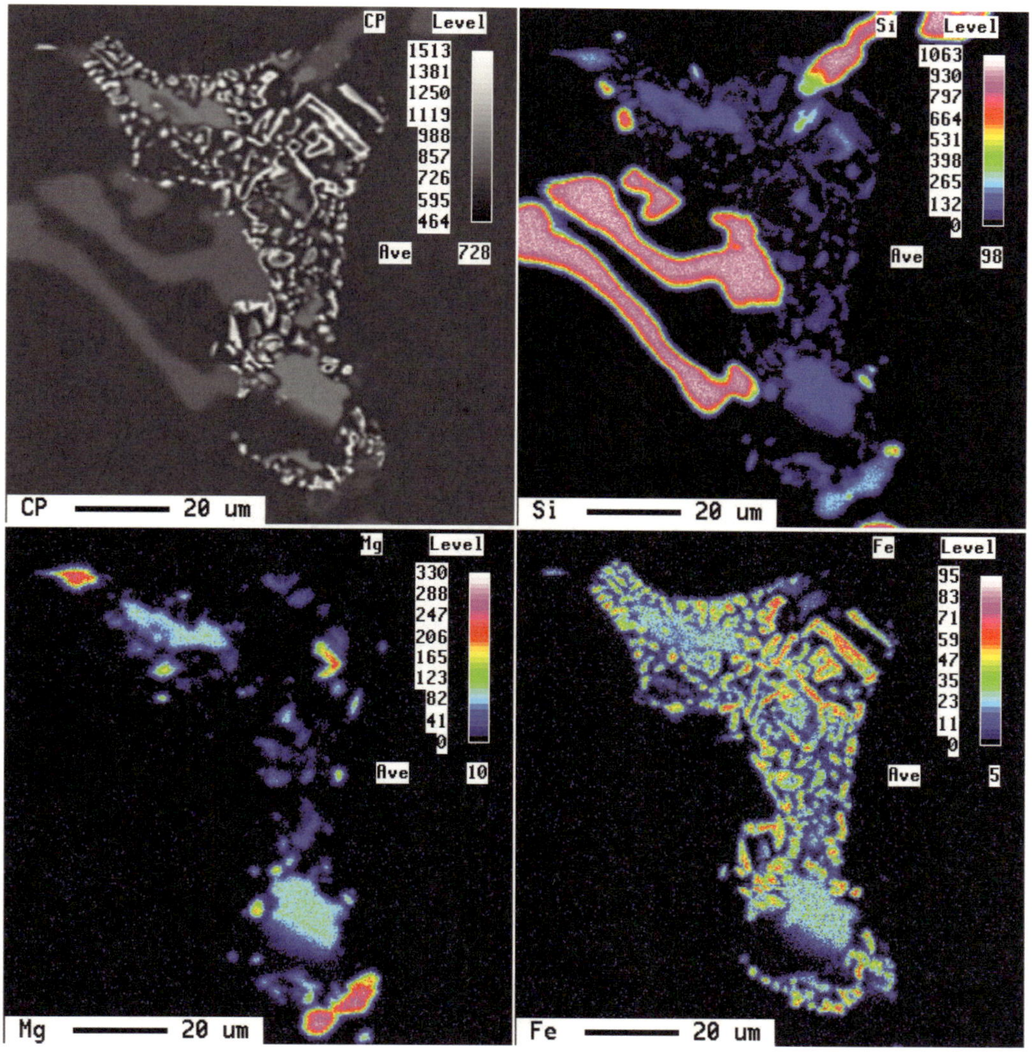

Figure 9. Decomposition of π-phase to β-phase during solutionizing at 540°C for 12 h [22].

The data was then treated using Minitab software and is presented in **Figures 10–12**.

Based upon the results presented in these figures, the following observations may be noted:

1. In the as-cast condition, increasing the Mg content leads to further transformation of the β-phase platelets to a Chinese script π-phase, regardless of the Fe content, thereby decreasing the harmful effect of the β-phase.
2. Increasing the solution heat treatment time leads to further decomposition of the π-phase, fragmentation of the β-phase, and spheroidization of the eutectic Si, resulting in an improvement in the alloy tensile properties.

3. Increasing the Fe level decreases the alloy ductility values, whereas the addition of Mg increases the % EF-values.

4. Introducing Be, Sr, or both, improves the alloy strength to some extent, regardless of the levels of Fe and Mg present.

5. Increasing solution treatment time from 5 to 12 h enhances the UTS and YS values.

6. Increasing the Mg-content results in improving the tensile properties; however increasing the iron levels markedly deteriorates these properties.

7. Additions of Be and Sr noticeably improves the mechanical properties of the alloys with the same Fe and/or Mg contents; however, these enhancements in mechanical properties are markedly observed at higher Mg contents and reduced levels of Fe.

Alloy code	**Property**	**As-cast**	**SHT 5 h**	**Aging time (h)—SHT 5 h**				
				2	**4**	**6**	**8**	**12**
A1	UTS (MPa)	204	262	342	358	361	354	360
	YS (MPa)	97	116	237	267	277	282	292
	El (%)	6.5	15.9	10	8.7	8.1	6.2	6.5
A3	UTS (MPa)	204	263	343	358	381	383	386
	YS (MPa)	114	141	254	292	330	333	347
	El (%)	3.4	7.2	5.96	4.3	3.1	2.8	2.0
C3	UTS (MPa)	201	239	320	328	366	366	360
	YS (MPa)	116	135	270	297	345	350	357
	El (%)	2.6	4.8	1.7	1.0	1.0	0.9	0.7
Alloy code	**Property**	**As-cast**	**SHT 12 h**	**Aging time (h)—SHT 12 h**				
				2	**4**	**6**	**8**	**12**
A1	UTS (MPa)	204	255	340	358	360	362	362
	YS (MPa)	97	107	229	273	284	274	280
	El (%)	6.5	17.6	12.3	10	8.7	9.5	8.5
A3	UTS (MPa)	204	287	351	368	382	382	384
	YS (MPa)	114	154	266	295	321	336	353
	El (%)	3.4	9.7	6.5	3.6	4.0	2.7	2.0
C3	UTS (MPa)	201	250	317	334	367	365	364
	YS (MPa)	116	152	294	301	317	326	327
	El (%)	2.6	4.3	1.0	1.0	0.9	0.7	0.7

Table 3. Tensile properties of reference alloys A1, A3, and C3 in the as-cast and heat-treated conditions.

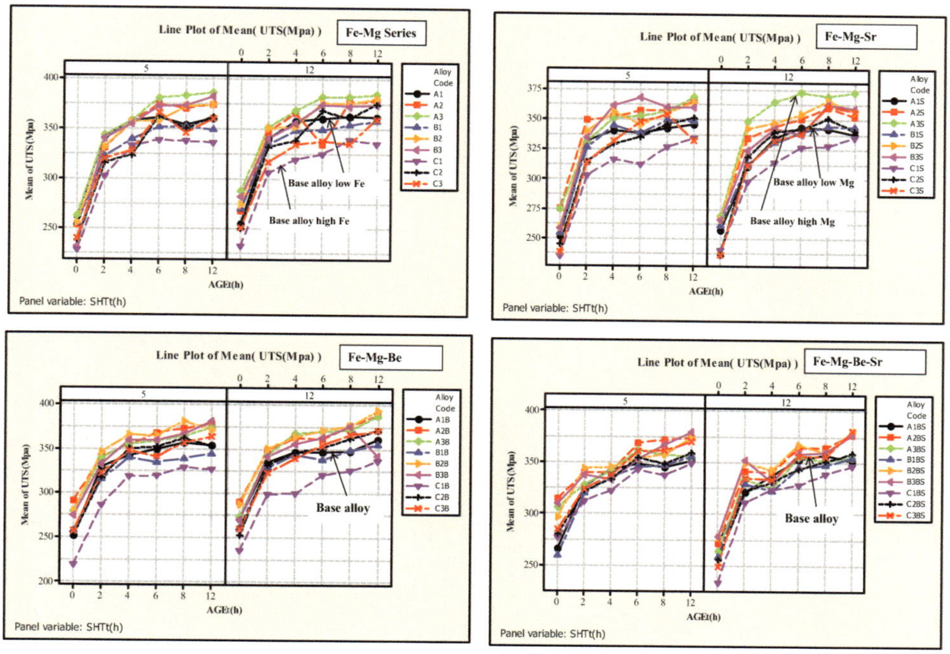

Figure 10. Variation in the UTS values as a function of alloying elements.

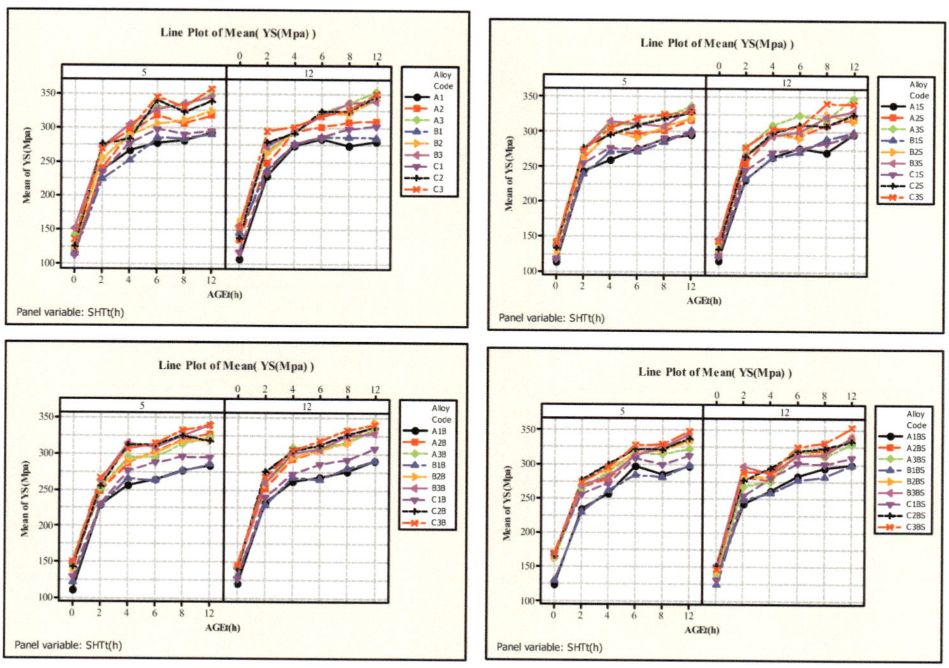

Figure 11. Variation in the YS values as a function of alloying elements.

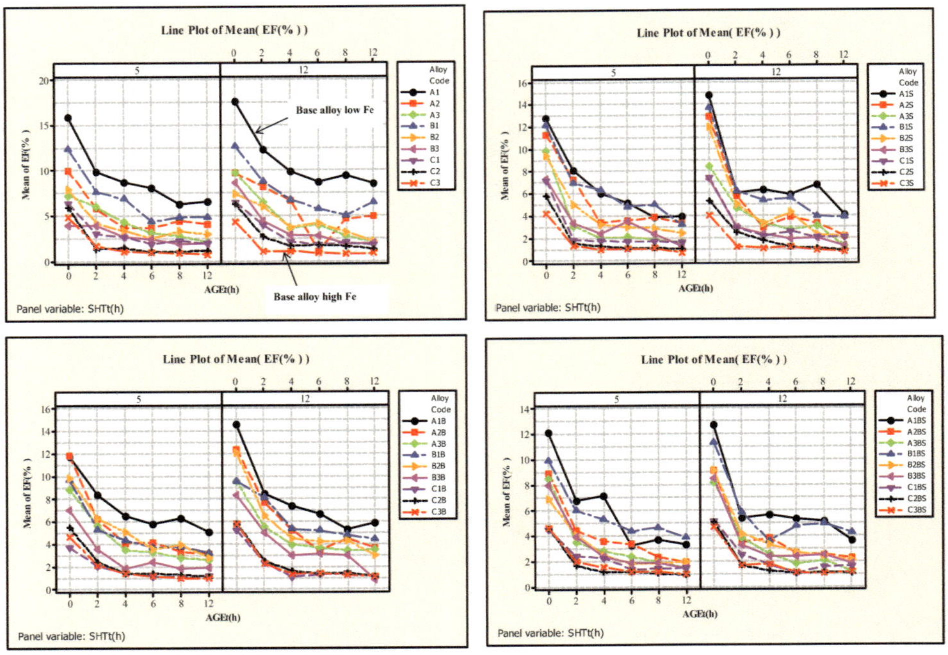

Figure 12. Variation in the % elongation to fracture (EF) values as a function of alloying elements.

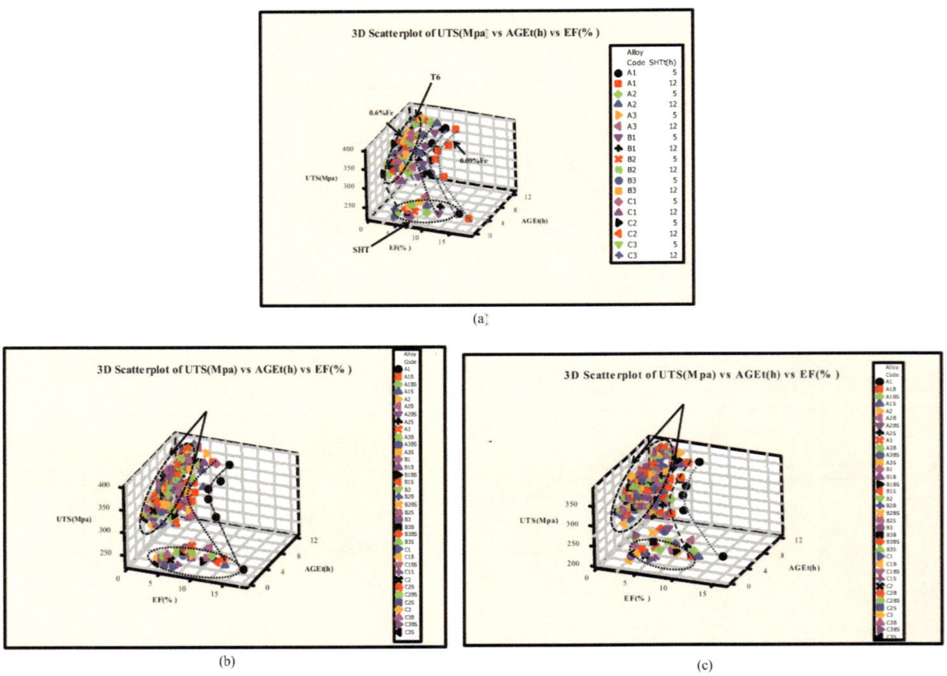

Figure 13. The 3D scattered plot of UTS versus %EF versus aging time: (a) base alloy, (b) all alloys in the T6 condition following solutionizing for 5 h at 540°C, and (c) all alloys in the T6 condition following solutionizing for 12 h at 540°C. The black arrows indicate the major axis of the oval circles.

It is evident from **Figure 13a** that Fe is the major element in determining the alloy performance. Introduction of Sr or Be or both would help in minimizing the harmful effect of Fe but to a limited extent. Increasing the solutionizing time to 12 h improved the alloy ductility due to coarsening of the eutectic Si particles. **Figure 13b** shows the UTS values of all the 36 alloys used in the present study for a solutionizing time of 5 h. It is clear that most of the T6 values are falling within a narrow zone represented by the oval circle moving the major axis to somewhat higher UTS levels with an increase in the UTS range (arrows). With increase in the solutionizing time from 5 to 12 h (**Figure 13c**), the width of the oval circle increased, accompanied by a slight drop in the major axis direction to lower UTS levels.

4. Concluding remarks

The present work was carried out on a series of heat-treatable aluminum-based aeronautical alloys containing various amounts of magnesium (Mg), iron (Fe), strontium (Sr), and beryllium (Be). From an analysis of the results obtained, the following concluding remarks may be made.

1. The addition of beryllium produces partial modification of the eutectic silicon particles in a manner similar to that reported for the addition of magnesium. The eutectic temperature was reduced by 10°C as a result of the addition of 0.8 wt% Mg. The solidification curves and their derivatives of Sr-free alloys with high Fe and Mg contents showed that there was a peak at 611°C consequent to the formation of a Be-Fe (Al_8Fe_2BeSi) phase, this peak was very close to the peak corresponding to the formation of α-Al. The morphology of the Be-Fe phase was recognized to be a script-like one.

2. Increasing the duration of solution heat treatment enhances the tensile properties of the alloys through the decomposition of the π-iron phase into the β-iron phase, fragmentation of the β-phase, and change in the geometry of both π-phase and silicon particles (i.e., spheroidization). Two mechanisms of eutectic Si particle coarsening during solution heat treatment were observed to occur: (1) Ostwald ripening in the solution and (2) clustering/collisions of the Si particles. Coarsening increases with increased solution heat treatment time.

3. Higher Mg contents are believed to be beneficial to the tensile properties (i.e., ductility, ultimate tensile strength, and yield strength) in that they oppose the detrimental effect of increasing the Fe content on the same tensile properties. Additions of Be and Sr noticeably improve the mechanical properties of the alloys with the same Fe and/or Mg contents; however, these enhancements in mechanical properties are markedly observed at higher Mg contents and reduced levels of Fe. In the case of high levels of Fe, addition of Be is preferable because it neutralizes the deleterious effects of Fe phases in cast aluminum-silicon alloys; though adding 500 ppm of Be, in case of high Fe contents, is inadequate to interact with other alloying elements. During the melting process, the possible existence of a Be-Sr phase (probably $SrBe_3O_4$) is believed to consume some of the Be content in the alloy, leading to both reduction in the free Be content as well as the alloy mechanical properties.

4. The alloy tensile properties are highly affected by varying the solutionizing and aging treatments times. In the 356 base alloy, the main strengthening effects were confirmed to be related to the formation of Mg_2Si precipitates. The improvements in the yield strength values are attributed to higher Mg contents, lower levels of Fe, addition of Be, Sr-modification, and solutionizing and aging times. Such enhancements in yield strength values are of great importance to the aeronautical industry because the design considerations in that field are mainly influenced by the yield strength.

Acknowledgements

The authors would like to thank Amal Samuel for enhancing the quality of the images and figures used in the present manuscript. They also wish to thank Dr. A.M. Samuel for careful proofreading of the manuscript.

Author details

Mohamed F. Ibrahim[1], Mohamed H. Abdelaziz[1], Herbert W. Doty[2], Salvador Valtierra[3] and Fawzy H. Samuel[1*]

*Address all correspondence to: fhsamuel@uqac.ca

1 Département des Sciences Appliquées, Université du Québec à Chicoutimi, Québec, Canada

2 General Motors, Materials Engineering, Pontiac, MI, USA

3 Nemak, S.A., Garza Garcia, N.L., Mexico

References

[1] Shivkumar S, Keller C, Trazzera M, Apelian D. Precipitation hardening in 356 alloys. In: Proceedings International Symposium on Production, Refining, Fabrication and Recycling of Light Metals; 26-30 August 1990; Hamilton, Ontario. 1990. pp. 264-278

[2] Apelian D, Shivkumar S, Sigworth G. Fundamental aspects of heat treatment of cast Al-Si-Mg alloys. AFS Transactions. 1989;**97**:727-742

[3] Edwards GA, Stiller K, Dunlop GL, Couper MJ. The precipitation sequence in Al-Mg-Si alloys. Acta Materialia. 1998;**46**(11):3893-3904

[4] Jorstad JL. Hypereutectic Al-Si casting alloys: 25 years, what's next? Silver Anniversary Paper. AFS Transactions. 1996;**104**:669-671

[5] Shivkumar S, Ricci S Jr, Steenhoff B, Apelian D, Sigworth G. An Experimental study to optimize the heat treatment of A356 alloy. AFS Transactions. 1989;**97**:791-810

[6] Yoshida KA, Arrowood RM. James. W. Evans, Microstructure and mechanical properties of A356 aluminum castings as related to various T6-Type heat treatments. In: Light Weight Alloys for Aerospace Applications III. Warrendale, PA: The Minerals, Metals, and Materials Society; 1995. p. 77-87

[7] Moustafa MA, Samuel FH, Doty HW. Effect of solution heat treatment and additives on the hardness, tensile properties and fracture behaviour of Al-Si (A413.1) automotive alloys. Journal of Materials Science. 2003;**38**:4523-4534

[8] Moustafa MA, Samuel FH, Doty HW, Valtierra S. Effect of Mg and Cu additions on the microstructural characteristics and tensile properties of Sr-modified Al-Si eutectic alloys. International Journal of Cast Metals Research. 2003;**15**:609-626

[9] Gilbert J, Leroy E. Aluminum Alloy Castings: Properties, Processes and Applications. ASM International, Materials Park, OH; 2004

[10] Barresi J, Kerr MJ, Wang H, Couper MJ. Effect of magnesium, iron and cooling rate on mechanical properties of Al-7Si-Mg foundry alloys. AFS Transactions. 2000;**114**:563-570

[11] Musmar S, Mucciardi F, Gruzleski J, Novwel A. In-situ thermal analysis technique for aluminum alloys 356, 319, Al-X Si, Al-Si-Cu X Mg, and 6063-experimental study. AFS Transaction. 2007;**115**:1-15

[12] Robison ST, Gonczy ST, Foley RD, Hartman K. Effect of different chill materials on aluminum casting solidification and mechanical properties. AFS Transactions. 2013;**121**:1-7

[13] Hatch JE. Aluminum: Properties and Physical Metallurgy. Metals Park, OH: American Society for Metals; 1984. p. 50-51

[14] Bäckerud SL, Chai G, Tamminen J. Solidification Characteristics of Aluminum Alloys. Vol. 2: Foundry Alloys. Oslo, Norway, AFS/Skanaluminium; 1990

[15] Taylor P. Ostwald ripening in emulsions. Advances in Colloid and Interface Science. 1998;**75**:107-163

[16] Samuel AM, Samuel FH, Doty HW. Observations on the formation of β-AlFeSi phase in 319 type Al-Si alloys. Journal of Materials Science. 1996;**31**:5529-5539

[17] Ibrahim MF, Samuel AM, Samuel FH, Ammar HR, Soliman M. On the impact toughness of Al-B4C MMC: the role of minor additives and heat treatment. In: 119th Metalcasting Congress, AFS 2015; 21-23 April 2015; Columbus, OH. 2015

[18] Ibrahim MF, Elgallad EM, Valtierra S, Doty HW, Samuel FH. Metallurgical parameters controlling the eutectic silicon charateristics in Be-treated Al-Si-Mg alloys. Materials. 2016;**9**(2):1-17

[19] Samuel AM, Doty HW, Valtierra S, Samuel FH. The effect of Bi-Sr and Ca-Sr interactions on the microstructure and tensile properties of Al-Si-based alloys. Materials. 2016;**9**(3):1-13

[20] Tavitas-Medrano FJ, Doty HW, Valtierra S, Samuel FH. On the enhancement of the impact toughness of A319 alloys: Role of Mg content and melt treatment. International Journal of Metalcasting. 2017;**11**(3):536-551

[21] Garza-Elizondo GH, Samuel AM, Valtierra S, Samuel FH. Effect of transition metals on the tensile properties of 354 alloy: Role of precipitation hardening. International Journal of Metalcasting. 2017;**11**(3):413-427

[22] Elsharkawi EA, Samuel FH, Simielli E, Sigworth GK. Influence of solutionizing time, modification, and cooling rate on the decomposition of Mg-containing iron intermetallic phase in 357 alloys. Transactions of the American Foundry Society. 2012;**120**:55-65 Paper No. 12-009

[23] Elsharkawi EA, Samuel E, Samuel AM, Samuel FH. Effects of Mg, Fe, Be additions and solution heat treatment on the pi-AlMgFeSi iron intermetallic phase in Al-7Si-Mg alloys. Journal of Materials Science. 2010;**45**(6):1528-1539

[24] Bailey WA, Bossing EN. High strength aluminum alloy airframe castings. AFS Transactions. 1961;**69**:660

[25] Fuoco R, Moreira M. Fatigue cracks in aluminum cylinder heads for diesel engines. International Journal of Metalcasting. 2010;**4**(4):19-32

[26] Maube SE, Wangombe DN, Maranga SM, Kihiu JM. Effect of cooling rate and heat treatment on the microstructure and impact resistance of recycled aluminum sand cast alloy. Proceeding of the Mechanical Engineering Conference. May 2012;**4**:214-218

[27] Juang SH, Wu SM. Study on mechanical properties of A356 alloys enhanced with preformed thixotropic structure. Journal of Marine Science and Technology. 2008;**16**(4): 271-274

[28] Tuttle MM, McLellan DL. Silicon particle characteristics in Al-Si-Mg castings. AFS Transactions. 1982;**90**:13-32

[29] Lados DA, Apelian D. Solution heat treatment effects on microstructure and mechanical properties of Al-(1 to 13Pct)Si-Mg cast alloys. Metallurgical and Materials Transactions B. 2011;**42B**:171-180

[30] Bailey WA. Beryllium effect on strength and mechanical properties of 356 variant—T6 aluminum alloys. AFS Transactions. 1964;**72**:443-454

Chapter 2

Rapid Solidification of Undercooled Melts

Xiaolong Xu, Hua Hou and Feng Liu

Additional information is available at the end of the chapter

http://dx.doi.org/10.5772/intechopen.70666

Abstract

Rapid solidification and microstructure evolution of deeply undercooled bulk concentrated Ni-20%at.Cu and Co-20%at.Pd alloys are strictly and systematically evaluated. First, thermodynamics of the undercooled melt is discussed. Consideration is provided for not only the systematic microstructure evolution within a broad undercooling range, but also the dendrite growth mechanism and the rapid solidification characteristics. The dendrite growth in the bulk undercooled melts was captured by a high speed camera. The first kind of grain refinement occurring in the low undercooling regimes was explained by a current grain refinement model. Besides for the dendrite melting mechanism, the stress originating from the solidification contraction and thermal strain in the first mushy zone during rapid solidification could be a main mechanism causing the second kind of grain refinement above the critical undercooling. This internal-stress led to the distortion and breakup of the primary dendrites and was semi-quantitatively described by a corrected stress accumulation model. It was found that the stress induced recrystallization could make the primary microstructures refine substantially after recalescence.

Keywords: rapid solidification, undercooling, grain refinement, alloys, stress, recrystallization

1. Introduction

Grain refinement is an interesting phenomenon that has an important scientific significance and has been numerously investigated [1–22]. In order to understand of grain refinement phenomenon better, numerous studies have been extensively executed and many different grain refinement mechanisms were suggested separately, such as dynamic nucleation [1], critical growth velocity [5, 6], kinetics induced growth instabilities [2–4], dendrite fragmentation [7, 9], dendrite remelting and [8, 9] recrystallization [7]. In 1959, Walker [1–8] had been firstly investigated the grain refinement event occurring in the rapid solidification of deeply undercooled pure Nickel melt. He found that the grain size would abruptly refine when the initial undercooling ΔT prior to nucleation exceeded a critical value. Thereafter, Powell [5] also found the existence of a critical undercooling when he investigated the microstructural evolution of Ag, Cu and Ni based alloys as a function of undercooling. Powell suggested that the recrystallization process during recalescence and post-recalescence periods would be responsible for the grain refinement at high undercooling ranges [5]. Herlach [6] measured the dendrite growth velocity in many undercooled melts, and he found that critical undercooling was correlated to a critical crystal growth velocity, which was about 20 m/s and was the solute diffusion velocity V_D in the undercooled melts. In the recent years, grain refinement occurring at undercoolings much lower than ΔT^* has attracted much attention [10–12]. During the post-recalescence period, the uniformly massive dendrite breakup will occur though the dendrite remelting. Karma [8] has proposed a physical mechanism which suggests that the break-up of dendrites due to remelting brings about the grain refinement both at low and high undercooling regimes. However, this model only considered the remelting due to liquid/solid interface tension, ignoring the effect of the chemical superheating [9] during recalescence, which may also play a much more important role in the dendrite remelting, and has a stronger influence on the final grain morphology. It is the main subject of the present paper to extend the current chemical superheating model to non-equilibrium solidification conditions and fully explain the dendrite remelting mechanism of the grain refinement events.

In the present study, we choose Ni-20at.%Cu alloy and Co-20at.%Pd alloy. The authors experimentally investigated the microstructural evolution of the Ni-20at.%Cu alloys as a function of initial undercooling and the physical mechanisms of the grain refinements occurring at low undercooling regimes. In combination with the current dendrite growth model, we theoretically analyzed the dendrite remelting in the undercooled alloys by an extended chemical superheating model for non-equilibrium solidification of undercooled binary single phase alloys.

2. Material and methods

Ni-20at.%Cu (atomic percent) alloy and Co-20at.%Pd (atomic percent) alloy samples each weighing about 3 g, were prepared by *in situ* melting pure Ni pieces (99.95% purity) and pure Cu pieces (99.9 wt.% purity), pure Co pieces (99.99% purity) and pure Pd pieces (99.99 wt.% purity), under the protection of argon (Ar) atmosphere in a vacuum chamber. Before melting, the surfaces of the metals were cleaned and grinded off mechanically to remove the surface oxide layer and were etched chemically in HCl solution diluted by alcohol. A high purity quartz crucible containing the alloy specimen was placed in the center of an induction coil. The melting process was conducted in the vacuum chamber. Undercooling experiment was executed by using of fluxing liquid and high frequency induction heating and thermal cycle under the protection of argon atmosphere. The vacuum chamber was evacuated and subsequently backfilled with 99.99% argon gas. Each sample was melted, superheated, solidified and subsequently remelted in superheating–cooling cycles to obtain various undercoolings. For each of the alloys, 20–30 undercoolings were made and a natural cooling rate of 20 K/s IS applied in these experiments. After the high frequency power source was turned off, the alloy sample was spontaneously cooled to room temperature, while the cooling curve of the specimen was monitored by an infrared pyrometer with an accuracy of 5 K and a response time of 10 ms.

3. Solidification process

In the present study, we used a high speed camera to capture the solidification process, i.e., to capture the heat releasing process upon solidification. It can be seen from **Figure 1** that for ΔT = 95 K (95°C), 160 K (160°C), 200 K (200°C) and 270 K (270°C), the nucleation of the undercooled melts all started from the interface between the B_2O_3 glass and the undercooled melt. However, for a higher enough undercooling, ΔT = 320 K (320°C), the start stage of heat releasing process was between the walls of the quartz crucible and the undercooled melts. It is generally considered that the higher of the extent of purification of an undercooled melt, the higher undercooling could be achieved. In addition, the impurities are mainly in the B_2O_3 melt and the interface between B_2O_3 melt and the undercooled melt. The present experiment phenomenon proves that the instabilities of the undercooled melts (i.e., nucleation) with ΔT = 95 K (95°C), 160 K (160°C), 200 K (200°C) and 270 K (270°C) are mainly due to the impurities absorbed in the interface between the B_2O_3 melt and the undercooled melt. However, for the higher undercooled alloy ΔT = 320 K (320°C), the impurities are rather less and the instabilities of the undercooled melt was in the interface between the quartz crucible and the

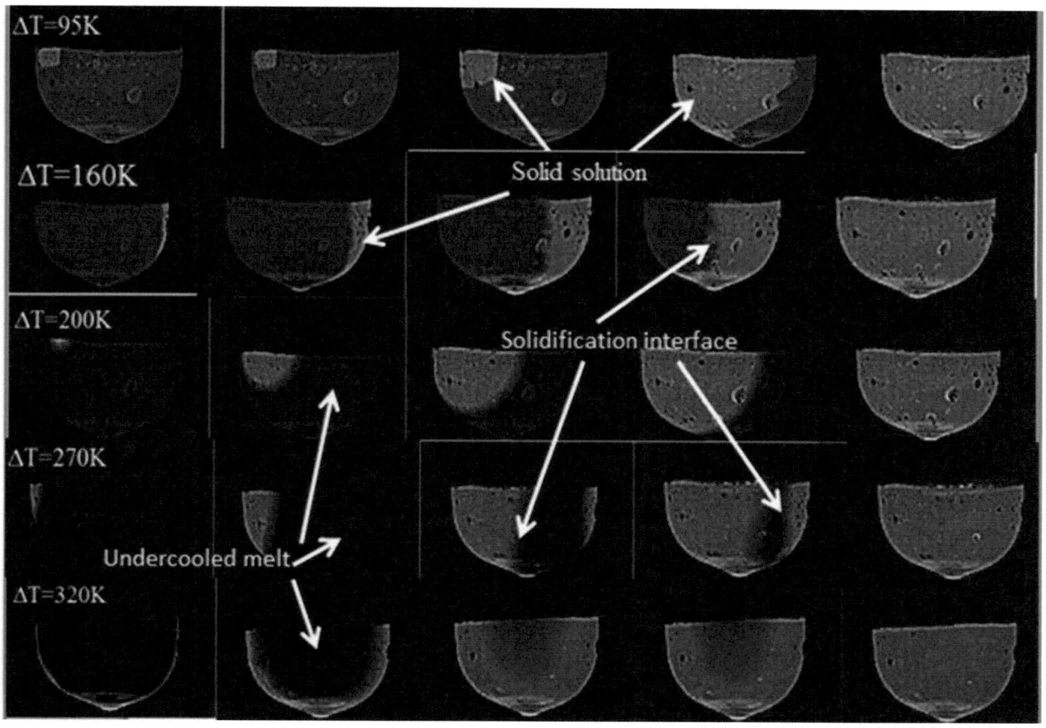

Figure 1. The high-speed video pictures of the undercooled Ni-20at.%Cu different undercoolings.

undercooled melt. Besides, it can be seen that multiple nucleation of solidification occurred. During rapid solidification of the undercooled melts, such as 95 K (95°C) and 160 K (160°C), it can be seen that the coarse primary dendrites are growing towards the undercooled residual melt. However, at high undercoolings such as 200 K (200°C), 270 K (270°C) and 320 K (320°C), much finer dendrites are growing into the melts. Thus, it can be concluded that the higher the undercooling of the melt, the finer the growing dendrites would be.

4. Dendrite growth analysis

Firstly, we need to use the BCT model by Boettinger, Coriell and Trividi [13] to calculate the solidification parameters. According to the models, the initial undercooling ΔT at the dendrite tip consists of four parts,

$$\Delta T = \Delta T_r + \Delta T_c + \Delta T_k + \Delta T_t \tag{1}$$

where ΔT_r, ΔT_k, ΔT_c, and ΔT_t are the thermal undercooling, solutal undercooling, curvature undercooling and interfacial kinetic undercooling, respectively. Details of the BCT model are referred to Ref. [13].

Using the physical parameters of the Ni-20at.%Cu alloy listed in **Table 1**, the undercooling components, the crystal growth velocity and the tip radius at the growing dendrite tip, can be calculated. The results are illustrated in **Figures 2** and **3**. It should be clarified that the table is original and we are using only data from Ref. [11].

It is well known that [13–15], in a single phase alloy, the condition of diffusional equilibrium is gradually becoming less important with the increase of solidification velocity and undercooling in front of the dendrite tip. Therefore, solute rejection is reduced and solutal undercooling decreases as the interface concentration approaches the melt composition. This ultimately causes partitionless solidification, which is solely controlled by thermal gradient. For Ni-20at.%Cu alloys (**Figure 2**), when the initial undercooling $\Delta T < 35$ K (35°C), the solute undercooling ΔT_c is larger than those of other undercooling components, so the dendrite growth in the undercooled melt is mainly controlled by the solute diffusion in front of the dendrite tip and the crystal growth velocity is very low, therefore, coarse dendrites can be resultant. As the undercooling continuously increases, the effect of thermal diffusion on the dendrite growth becomes strong. In the undercooling range of about 35 K (35°C) < ΔT < 65 K (65°C), ΔT_t starts to exceed ΔT_c, which reaches a maximum at $\Delta T = 55$ K (55°C) where the dendrite tip also reaches its maximum and is the most unstable due to the maximum solute undercooling. In this undercooling range, the action of solute undercooling ΔT_c and thermal undercooling ΔT_t is on the same level. When $\Delta T > 70$ K (70°C), ΔT_t increases rapidly and exceeds ΔT_c substantially, the dendrite tip radius accordingly decreases (**Figure 3**) and the crystal growth rate V increases rapidly (**Figure 3**). The increasing action of thermal diffusion results in directionality of dendrites. Solute diffusion is replaced by the thermal diffusion to predominantly control the dendrite growth process, which indicates a transition from the equilibrium of a solidification controlled by the solutal gradient to a thermally controlled growth process owing to a relaxation of diffusional equilibrium at the

http://dx.doi.org/10.5772/intechopen.70666

Parameters	Value of Ni-20at.%Cu alloys
Heat of fusion ΔH_f/kJ mol^{-1}	17,160
Specific heat of the liquid C_p/J^{-1} mol^{-1} K^{-1}	38.5
Dynamic viscosity of the liquid μ/Pas	10^{-3}
Molar volume of the liquid V_m^L/m^3 mol^{-1}	8.06×10^{-6}
Molar volume of the solid V_m^S/m^3 mol^{-1}	7.08×10^{-6}
Liquidus T_L/K	1680 K (1407°C)
Solute diffusivity in the liquid D_L/m^2 s^{-1}	3×10^{-9}
Thermal diffusivity α_L/m^2 s^{-1}	7×10^{-6}
Atomic space a_0/m	4×10^{-10}
Interfacial energy $\sigma_{S,L}$/J m^{-2}	0.29
Speed of sound in the melt V_0/m s^{-1}	4000
Equilibrium liquidus slope m_L/K (at%)$^{-1}$	−2.525
Equilibrium solute partition coefficient k_0	0.724
Solidification time t_f/s	0.1
Size of the mushy zone a/m	0.01
Gibbs–Thomson coefficient Γ	3.25×10^{-7}
Solid fraction at the dendrite coherency point f_S^{coh}	0.15
Solidification shrinkage of the primary phase β	0.1215
Atomic diffusive speed V_D/m s^{-1}	20

Table 1. Physical parameters used in the calculation [9].

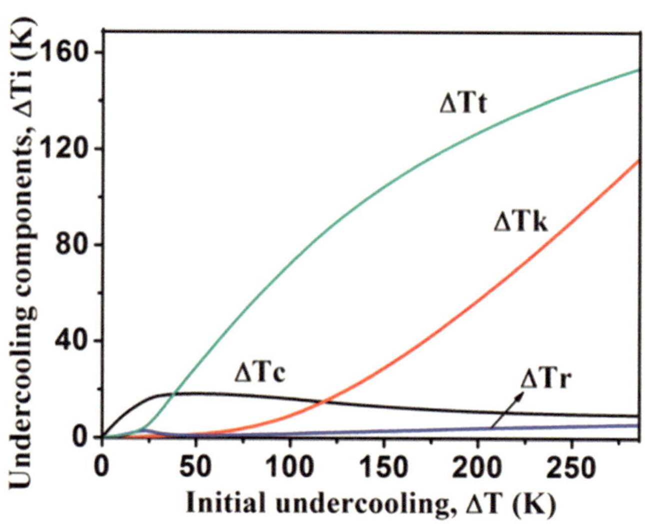

Figure 2. The undercooling constituents as a function of initial undercoolings for Ni-20at.%Cu alloys.

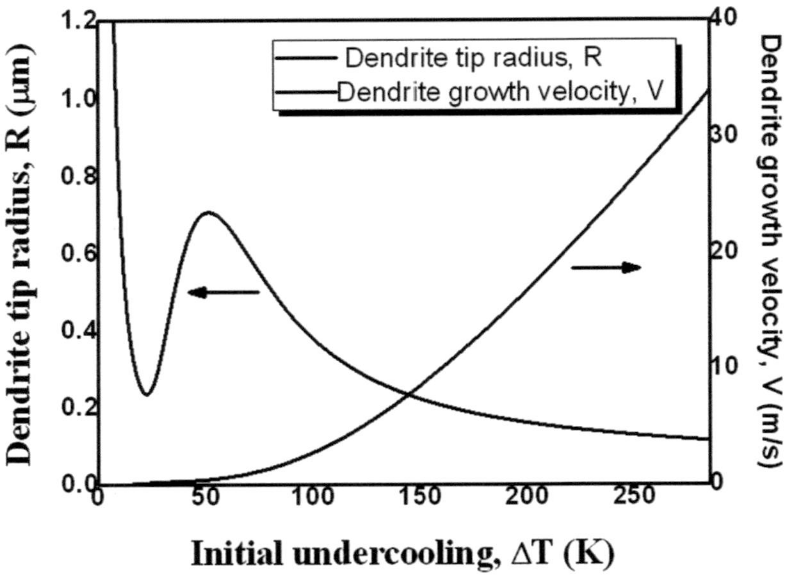

Figure 3. The variation of dendrite tip radius and growth velocity as a function of initial undercoolings for Ni-20at.%Cu alloys.

solid/liquid interface. Therefore, the higher the initial undercooling and the thermal undercooling, the more convenient it becomes for dissipating of the latent heat at the dendrite tip during the crystal growth.

5. Structure evolution of undercooled alloys

As the increasing of undercooling, the Ni-20%at.Cu alloys undergo two kinds of grain refinements: one taking place at low undercoolings, and the other occurring at high undercoolings. The grain sizes and the corresponding typical microstructures at various undercoolings are shown in **Figure 4**. Pioneer investigations [1–11] have shown that as the initial undercooling increases, the microstructures of the deeply undercooled binary single phase solid solution alloys evolves as: coarse dendrites → fine equiaxed grains →coarse dendrites plus fine equiaxed grains → fine equiaxed grains. In the present investigation, we found that the microstructural evolution of the deeply undercooled Ni-20%at.Cu alloys was basically follows the above law. Subjected to small undercoolings, e.g., at $\Delta T < \Delta T_1 = 45$ K, the solidification microstructures were coarse dendrites (**Figure 4a** and **b**) with highly developed secondary dendritic arms. As increasing ΔT, coarse dendrites gradually refined into granular grains by remelting. For the samples nucleated at undercooling about ΔT_1 – $\Delta T_2 = 90$ K, the overall solidification microstructures are occupied by fine grains (**Figure 4c**) with a few dendrite skeletons, as shown in the black ellipse marked in **Figure 4c**. A further increase of the undercooling leads to the rising of the grain size again. When $\Delta T > \Delta T_3 = 110$ K, much finer dendrite beams form because of that the controlling mechanism of dendrite

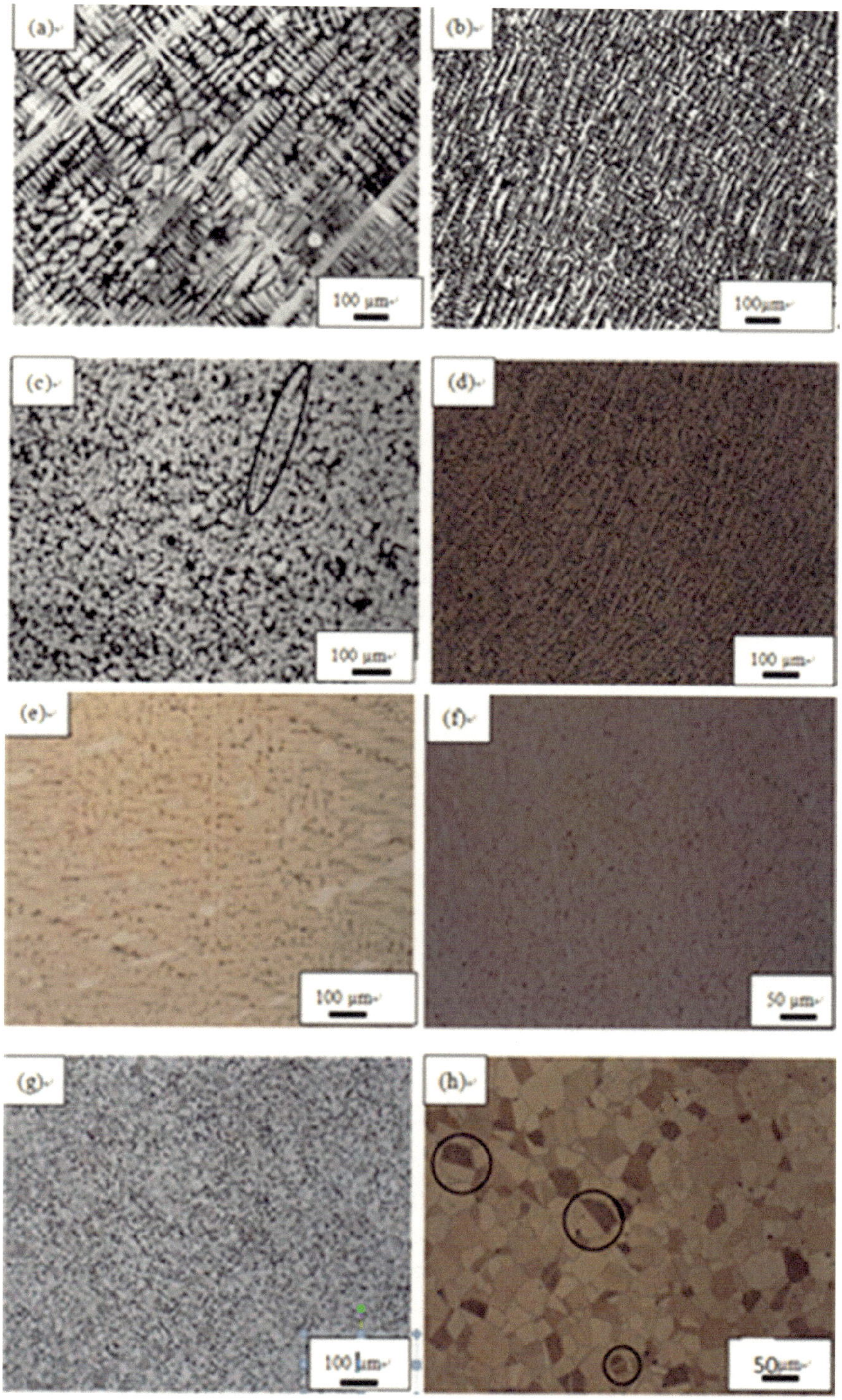

Figure 4. As-solidified microstructures of Ni-20at.%Cu alloy at small undercoolings: (a) ΔT = 17 K; (b) ΔT = 30 K; (c) ΔT = 72 K; (d) ΔT = 107 K; (e) ΔT = 158 K; (f) ΔT = 170 K; (g) ΔT = 182 K; (h) ΔT = 261 K.

growth has transformed from solute diffusion to thermal diffusion in the melt ahead of the dendrite tip. Here, ΔT_3 is defined as the characteristic undercooling at which there are no obvious dendrite fragments in the microstructures, and only large dendritic crystals are found (**Figure 4d** and **e**). At undercoolings larger than the critical value $\Delta T^* = 170$ K, the overall microstructures are refined again (**Figure 4f** and **g**). As ΔT exceeds 182 K, the microstructures consist of completely fine equiaxed grains and there are many twins in the grains, see the black circles in **Figure 4h**. The formation of annealing twins in fcc metals and alloys are due to the growth accidents on {1 1 1} propagating steps present on migrating grain boundaries. As a result of the accidents, Shockley partial dislocations are generated contiguously to boundaries. These partial dislocations repel each other and glide away from the boundary to produce a twin.

When an alloy melt solidifies at a small undercooling, e.g., $\Delta T = 50$ K in the case of undercooled Co-20%at.Pd, the advancement of solid/liquid (S/L) interface is controlled by solute diffusion ahead of the S/L interface. Solidification of the melt proceeds rather slowly and results in the formation of well-developed dendritic microstructures (see **Figure 5a**). With increasing undercooling, a grain refinement event is observed in the undercooled Co-20%at.Pd alloy at 50–265 K. The first grain refinement in an undercooled single phase alloy is generally believed to be caused by the remelting effect during the post-recalescence period. When ΔT reaches the range of 265–280 K, which is above the hypercooling limit, due to a zero remelting effect, the dendrite trunks cannot be remelted and therefore can be preserved in the solidification microstructure, leading to the presence of the elongated grained structure. The appearance of elongated grained structure in the hypercooled Co-20%at.Pd alloy is apparently ascribed to the vanished remelting effect in the post-recalescence period. As ΔT is higher than 280 K, the second grain refinement event occurs. The equiaxed grains with rather straight grain boundaries (see **Figure 5e** and **f**) strongly suggest that the formation of the solidification microstructure is due to recrystallization. The recrystallized microstructures in hypercooled alloys have been observed by Willnecker et al. in Co-20%at.Pd alloy and by Lu et al. in Ni75Pd25 alloy [12, 14]. In the present work, the occurrence of recrystallization in the hypercooled Co-20%at.Pd alloy with $\Delta T = 280$ K was considered to be induced by the accumulation of internal strain/stress caused by the dramatic shrinkage stress resulting from rapid solidification. For rapid solidification taking place at a rather high undercooling, the advance of the solid/liquid interface proceeds extremely fast. As a consequence, a large volume contraction owing to rapid solidification will cause a high shrinkage stress developing in the system.

Three typical temperature profiles corresponding to the solidification of Co-20%at.Pd alloy melts in different undercooling regimes are shown in **Figure 6**. It was found that when $\Delta T = 265$ K, the post-recalescence period presents in the temperature profiles, as seen in the plateau marked by the plateau duration time (Δt_{pl}) in **Figure 3a**. The plateau duration time Δt_{pl} is the time required for the residual liquid to transform into solid after recalescence. Δt_{pl} can be obtained by subtracting the time point of recalescence from the time point at which the first derivative of cooling curve after recalescence abruptly changes. When ΔT reaches 265 K, the

http://dx.doi.org/10.5772/intechopen.70666

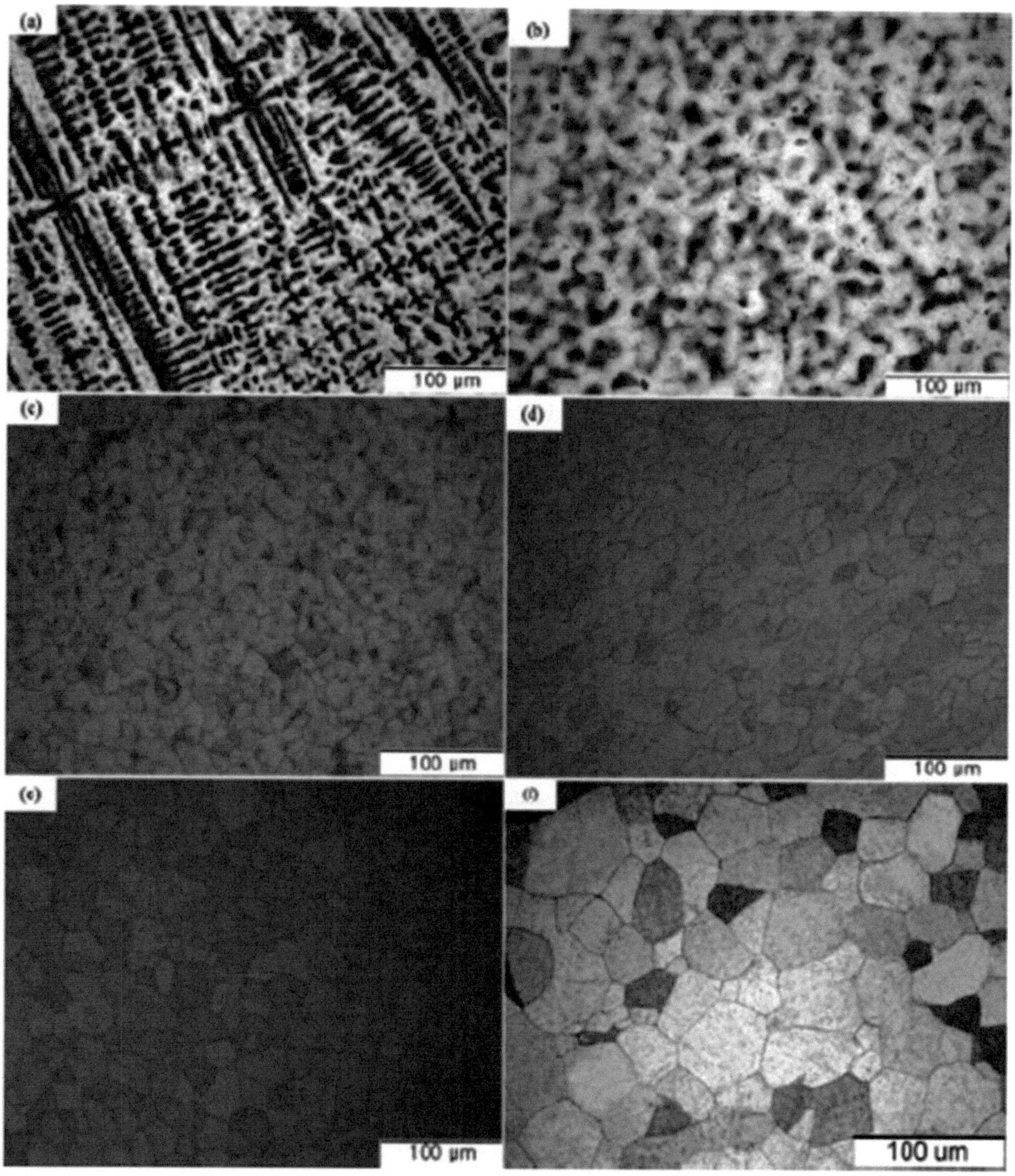

Figure 5. As-solidified microstructures of Co-20at.%Pd alloys undercooled by (a) 20 K, (b) 85 K, (c) 160 K, (d) 265 K, (e) 290 K, and (f) 340 K.

post-recalescence period does not appear, and the maximum recalescence temperature is lower than the solidus temperature of this alloy (see **Figure 3b** and **c**). Since the post-recalescence period will only vanish when the hypercooling limit of a metal is reached, the hypercooling limit of the Co-20%at.Pd alloy can be determined as 265 K.

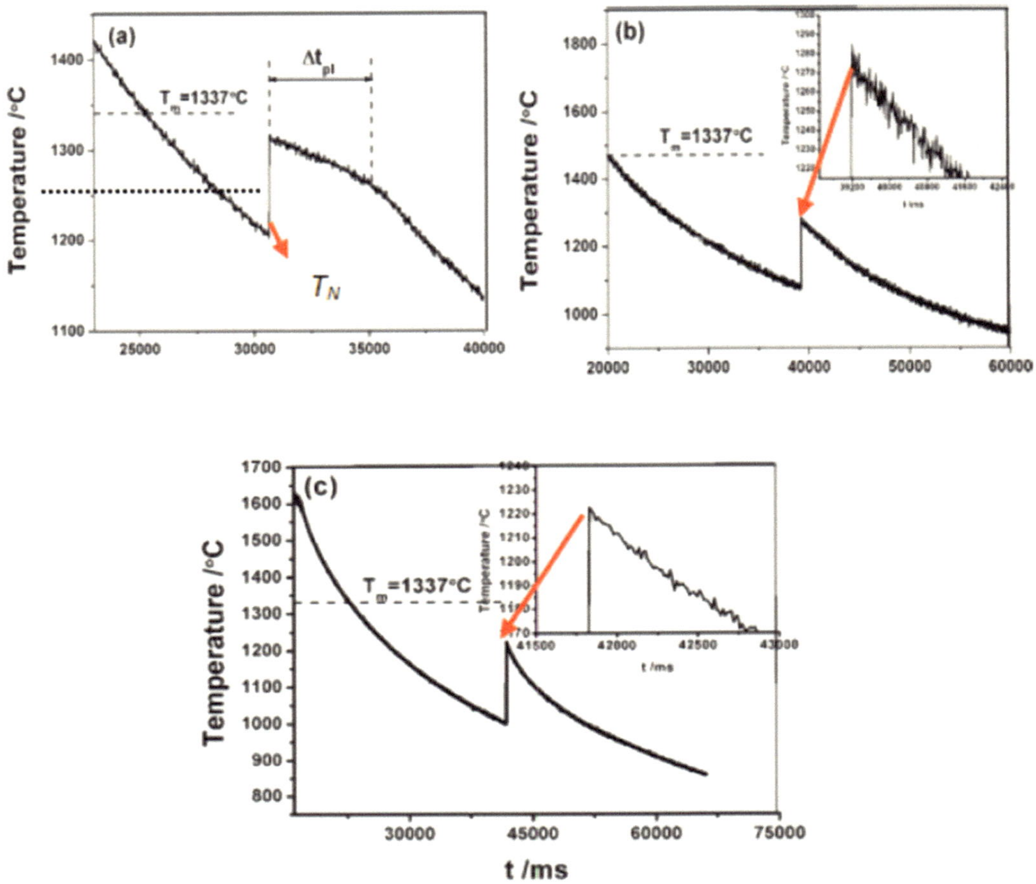

Figure 6. The measured average grain size as a function of initial undercooling of (a) Ni-20at.%Cu and (b) Co-20at.%Pd alloys.

6. Grain refinement mechanism of first refinement

The equilibrium phase diagram shows that the molten fraction of a solid alloy that has been heated into the solid-liquid binary phase regime is related to the chemical superheating [9]. Here, due to the non-equilibrium effect involved in rapid solidification of undercooled melts, it is reasonable to speculate that the non-equilibrium kinetic phase diagram can deal with the remelted fraction of the solid phase more realistically and practically. Then, extension of Li's model [9] will give a non-equilibrium chemical superheating model that can predict the remelted dendrite fraction by the following equation:

$$f_L = \frac{k(T_R - T_S)/\Delta T_0}{1 - (1-k)(T_R - T_S)/\Delta T_0} \tag{2}$$

This model is similar to Li's model [9], however it is an extended chemical superheating model considering the relaxation effect in undercooled liquid phase. In this extended model, k is the non-equilibrium solute partition coefficient. T_R is the maximum recalescence temperature corresponding to initial undercooling ΔT, T_S is the non-equilibrium solidus temperature corresponding to the composition C_S of the central part in the dendrite stem, and ΔT_0 is the non-equilibrium crystallization temperature range of the alloy with composition C_S. The value of f_L shows the extent of the remelting induced breakup of the dendrites, so f_L can be used to evaluate the tendency of remelting.

The remelted fraction of the primary dendrite is shown in **Figure 4**. The obvious difference between Li's model and the present model is that the present model incorporates local non-equilibrium effect (i.e., the relaxation effect in the undercooled bulk liquid phase) in the model derivation. Thus, it can be further inferred that the remelted fraction of the primary dendrite of the present model is mainly influenced by the composition of liquid phase of the present model, in which the relaxation effect plays an important role [19]. In the present experiments, the first kind of grain refinement of rapidly solidified Ni-20%at.Cu alloy emerged in the undercooling range of about 40 K (ΔT_2) to 90 K (ΔT_3). In consistency with the prediction of the present extended chemical superheating model, the first kind of grain refinement found in the experiments lies in the undercooling range which has very strong remelting tendency of dendrites (**Figure 7**). Comparing the predicted results of the Li's model (local equilibrium condition) and the present model (local non-equilibrium condition), we can see that under low undercooling ranges, these two model predict similar results. Both model (**Figure 7**)

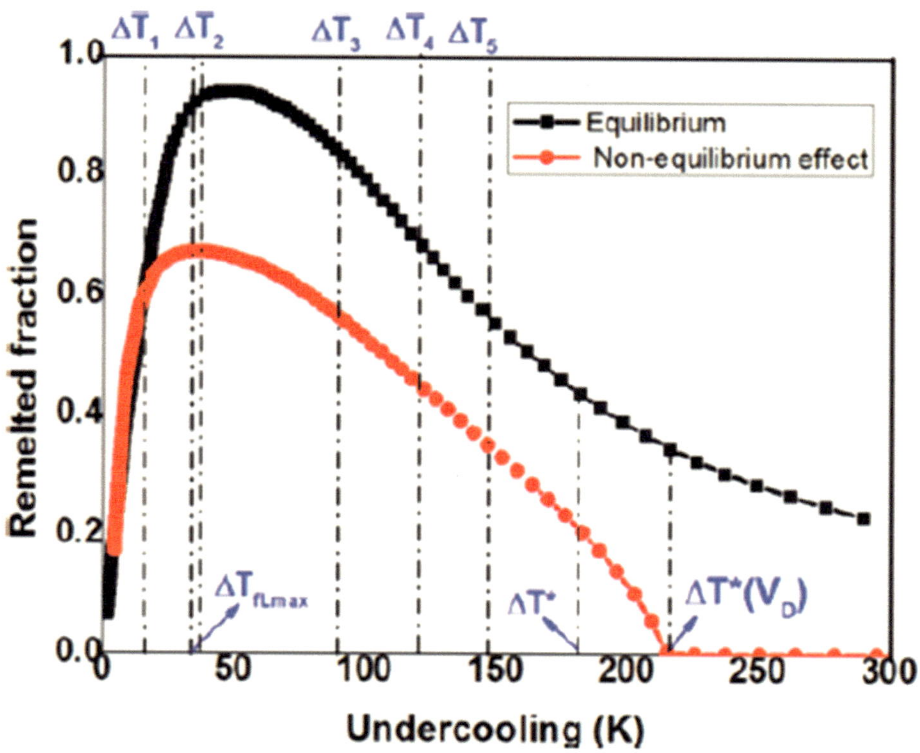

Figure 7. Remelted fraction of the primary dendrite at the highest recalescence temperature point in highly undercooled Ni-20at.%Cu.

predicts that the undercooling range of about 40–90 K has a strong dendrite remelting tendency, thus in this range the dendrites universally break up due to chemical superheating effect. While at high undercooling range severe discrepancy happens between these two models. As can be seen from **Figure 7** that, when the undercooling exceeds the critical undercooling ΔT^*, the discrepancy becomes more and more severe and achieves a maximum value at the undercooling value $\Delta T^*(V_D)$, i.e., the undercooling where the dendrite growth velocity exceeds the solute diffusion speed V_D in the undercooled liquid phase. Besides, at the undercooling value $\Delta T^*(V_D)$, the remelted fraction reaches zero and no chemical superheating induced dendrite remelting will happen. Thus, as previous study [10–18] showing, the grain refinement above the second critical undercooling ΔT^* is induced by the mechanisms of stress-induced breakup and recrystallization of the rapidly solidified dendrites, rather than dendrite remelting induced by chemical superheating, which can hardly affect the dendrite breakup in the high undercooling range as shown in **Figure 7**. Obviously, the significance of the present extended chemical superheating model is that it can predict good results in consistency with the experiment observation and the relaxation effect on non-equilibrium solidification of undercooled melt is clearly illustrated. The first grain refinement mechanism is also the chemical superheating mechanism and will not be given here for short. It has been revealed that the stress induced dendrite breakup and recrystallization process mainly occur during and after recalescence period. These two physical processes should be responsible for the grain refinement at high undercooling regimes [9].

7. Stress induced recrystallization mechanism ($\Delta T > \Delta T^*$)

Many researches [7–9, 15–17] have revealed that the stress induced recrystallization process mainly during recalescence period should be responsible for the grain refinement at high undercooling range. However, few strong and direct evidences have been found to support this speculation. We also have revealed this phenomenon in hypercooled Co-20at.%Pd alloys and found strong and direct evidences, see Ref. [13].

When the fraction of the solid phase exceeds about 10% during recalescence, the coherent dendrite networks are usually established [7, 9]. During rapid recalescence, as the dendrite fraction increases, the inter-dendritic permeability correspondingly decreases [7, 9]. Consequently the pressure gradient in the first mushy zone (FMZ) increases. The pressure gradient induces the inter-dendritic flow of liquid phase and therefore stress accumulated in the solid phase in the FMZ [9]. Considering inter-dendritic flow of melt induced by both solidification contraction and thermal strain, we will build a corrected stress accumulation model for rapid solidification of an undercooled melt was established. According to Ref. [9], the liquid volume flux $U(x)$ among dendrites can be expressed as:

$$U\left(\vec{x}\right) = \frac{\left|\vec{a}\right|}{t_f}\left[\beta_s\left(f_R^s - f_s\left(\vec{x}\right)\right)\right] \tag{3}$$

where $\vec{a}$ is the size of FMZ in which primary growth or recalescence occurs, t_f is the total solidification time of the primary phase, i.e., the recalescence time, $\beta_s=(\rho_l-\rho_s)/\rho_s$ is the solidification shrinkage of the primary phase, ρ_l and ρ_s are densities of liquid and solid phases respectively. f_R^s is the fraction of solid at the maximum recalescence temperature, f_s is the solid fraction during recalescence and $\vec{x}$ is the local position in the One-dimensional space coordinates.

The thermal strain induced inter-dendritic flow of liquid phase $U'(\vec{x})$ in the mushy zone can be expressed as [22]:

$$U'\left(\vec{x}\right) = -(1+\beta_s)\dot{E}\left(\vec{x}\right) \tag{4}$$

where $\dot{E}(x)$ is the accumulated strain rate in the dendrites and can be defined as [16]:

$$\dot{E}\left(\vec{x}\right) = \int f_s \dot{\varepsilon}\left(\vec{x}\right) d\,\vec{x} \tag{5}$$

where $\dot{\varepsilon}(\vec{x}) = \nabla v_{\vec{x}}(\vec{x})$ is the strain rate in the dendrites and $\dot{\varepsilon}(\vec{a}) = 0$.

According to Eq. (6) in Ref. [7] and Eq. (4), the total volume flow of molten feeding liquid is:

$$U_{total}\left(\vec{x}\right) = \frac{\left|\vec{a}\right|}{t_f}\left[\beta_s\left(f_R^s - f_s\left(\vec{x}\right)\right)\right] + (1+\beta_s)\int f_s\left(\vec{x}\right)\dot{\varepsilon}\left(\vec{x}\right) d\,\vec{x} \tag{6}$$

According to Darcy's equation, as demonstrated in Ref. [22], the pressure gradient in the liquid phase can be expressed as [16]:

$$-\nabla P_l\left(\vec{x}\right) = \frac{\mu}{K} U_{total}\left(\vec{x}\right) \tag{7}$$

where P_l is the pressure in the liquid phase during solidification, μ the dynamic viscosity and K the permeability of the two phase region as demonstrated in Ref. 17. According to the Carman-Kozeny relation demonstrated in Ref. 17, K can be expressed as [16]:

$$K = \frac{1}{5S_V^{s-l}\left(\vec{x}\right)^2} \frac{f_l\left(\vec{x}\right)^3}{f_s\left(\vec{x}\right)^2} = \frac{\lambda_2^2\left(\vec{x}\right) f_l\left(\vec{x}\right)^3}{80\, f_s\left(\vec{x}\right)} \tag{8}$$

where $G_L(\vec{x})$ is the temperature gradient ahead of the liquid/solid interface. $S_V^{s-l}(\vec{x})$ is.

the area of the dendrites in unit volume and $S_V^{s-l}(\vec{x}) = 4/\left(\sqrt{f_s(\vec{x})}\lambda_2(\vec{x})\right)$, $\lambda_2(\vec{x})$ is the spacing between secondary dendrite arms [23]. Applying linear micro-segregation model, as demonstrated in Ref. [23]:

$$dT/df_l\left(\vec{x}\right) = \Delta T_0 \tag{9}$$

where ΔT_0 is the solidification interval of the alloy. Using Eqs. (7)–(9) in combination with Eq. (12) in Ref. [7], the pressure of the mushy zone can be expressed as:

$$P(\vec{x}) = P_0 - \frac{160\mu\cdot\left|\vec{a}\right|^2\cdot\beta_s}{\left(f_R^s\right)^2 t_f \lambda_2^2}\left[f_s(\vec{x}) - f_{coh}(\vec{x}) + \frac{1}{1-f_s(\vec{x})} - \frac{1}{1-f_{coh}(\vec{x})} + 2\ln\frac{1-f_s(\vec{x})}{1-f_{coh}(\vec{x})}\right]$$
$$+\left\{\frac{320(1+\beta_s)\cdot\mu\cdot\left|\vec{a}\right|\cdot\Delta T_0\cdot\dot{\varepsilon}}{3G_S\cdot\left(f_R^s\right)^2\cdot\lambda_2^2}\int_{f_{coh}}^{f_s}\frac{f_s(\vec{x})}{f_l(\vec{x})^3}\left[f_s(\vec{x})^3 - f_{coh}(\vec{x})^3\right]df_s(\vec{x})\right\} \tag{10}$$

where f_{coh} is the fraction solid at the dendrite coherency point. The second term in Eq. (10) considers the liquid flow induced by thermal strain during rapid solidification. We can describe the thermal strain as:

$$\varepsilon = \alpha_{ther}(T - T_s) \tag{11}$$

where α_{ther} is a coefficient of thermal expansion. T_s is solidus temperature. Differentiating Eq. (10), we obtain thermal strain rate:

$$\dot{\varepsilon} = \alpha_{ther}\dot{T} \tag{12}$$

We define average thermal strain as:

$$\bar{\dot{\varepsilon}} = \alpha_{ther}\frac{\Delta T}{\Delta t} \tag{13}$$

In order to estimate the accumulated stress in the dendrites, we consider the one dimensional case. Then the momentum equation in solid phase is, as demonstrated in Ref. [24]:

$$0 = \nabla\left[f_s(\vec{x})\sigma_s(\vec{x})\right] + M_s^{\tau}(\vec{x}) \tag{14}$$

where $\sigma_s(\vec{x})$ is the x component of the stress in the solid phase and M_s^{τ} is the interfacial stress due to interaction with the liquid phase. The interfacial stress exerted upon the solid network from flowing liquid phase can be given as, as demonstrated in Ref. [24]:

$$M_s^{\tau}(\vec{x}) = P_l(\vec{x})\nabla f_s(\vec{x}) + M_s^d(\vec{x}) \tag{15}$$

where $M_s^d(\vec{x})$ is the dissipation of stress of interface and can be expressed by Darcy Eq., as demonstrated in Ref. [24]:

$$M_s^d(\vec{x}) = \frac{f_l(\vec{x})\mu}{K}U_{total}(\vec{x}) = -f_l(\vec{x})\nabla P_l(\vec{x}) \tag{16}$$

In combination with Eq. (16), the stress in the dendrite coherency obeys the following equation [24]:

$$-\nabla\left[f_s(\vec{x})\sigma_s(\vec{x})\right] = P_l(\vec{x})\nabla f_s(\vec{x}) - f_l(\vec{x})\nabla P_l(\vec{x}) = -\nabla\left[P_l(\vec{x})f_l(\vec{x})\right] \quad (17)$$

At the start of the coherent part of the first mushy zone, which is abbreviated as FMZ, the solid stress is equal to the hydrostatic pressure, i.e., $\sigma_s = -P_0$, where P_0 is the pressure in the liquid phase when the mushy zone starts to form. $f_s(\vec{x})$ is the solid fraction of the dendrite coherency. Integrating Eq. (10), we can obtain:

$$\sigma_S\left[f_s(\vec{x})\right] = \frac{1}{f_s(\vec{x})}\left[-P_0 + P_l(\vec{x})f_l(\vec{x})\right] = -\left[P_0 + \frac{f_l(\vec{x})}{f_s(\vec{x})}\Delta P_l(\vec{x})\right] \quad (18)$$

where $\Delta P_l = P_0 - P_l$ is the pressure drop in the liquid phase. Using Eqs. (10), (13) and (18), the stress resulted from inter-dendritic liquid flow induced by both solidification shrinkage and thermal strain is:

$$\sigma_S = \frac{160\mu\cdot|\vec{a}|}{(f_R^s)^2 t_f \lambda_2^2} \times \frac{\left(1-f_s(\vec{x})\right)}{f_s(\vec{x})}\left\{|\vec{a}|\cdot\beta_s\left[\begin{matrix} f_s(\vec{x}) - f_{coh} + \dfrac{1}{(1-f_s)} - \dfrac{1}{(1-f_{coh})} \\ +2\ln\left[\dfrac{1-f_s(\vec{x})}{1-f_{coh}}\right]\end{matrix}\right]\right.$$

$$\left. - \frac{2(1+\beta_s)\alpha_{ther}\Delta T_0 \Delta T}{3G_S} Q\left[f_s(\vec{x})\right]\right\} \quad (19)$$

where

$$Q\left[f_s(\vec{x})\right] = \int_{f_{coh}}^{f_s} \frac{f_s(\vec{x})}{f_l(\vec{x})^3}\left[f_s(\vec{x})^3 - f_{coh}(\vec{x})^3\right] df_s(\vec{x}) \quad (20)$$

Considering inter-dendritic flow of liquid induced by both solidification contraction and thermal strain, an analytic model of stress accumulation during rapid solidification of undercooled melts was established in the present work. According to this model, the stress accumulated in the dendrite coherency can be calculated by Eq. (19). G_S is temperature gradient of solid phase, respectively. In the current work, it is assumed that $\lambda_2 = \alpha R_{tip}$. Using the physical parameters listed in **Table 1**, σ_s as a function of ΔT can be calculated by Eqs. (19) and (20), as shown in **Figure 8**.

It can be seen from **Figure 8** that, with increasing ΔT, σ_s continuously increases. Once σ_s exceeds the yielding strength of the alloys, the dendrite networks will be plastically deformed, leading to dense dislocations and annealing twins and a storage of micro-strain in the

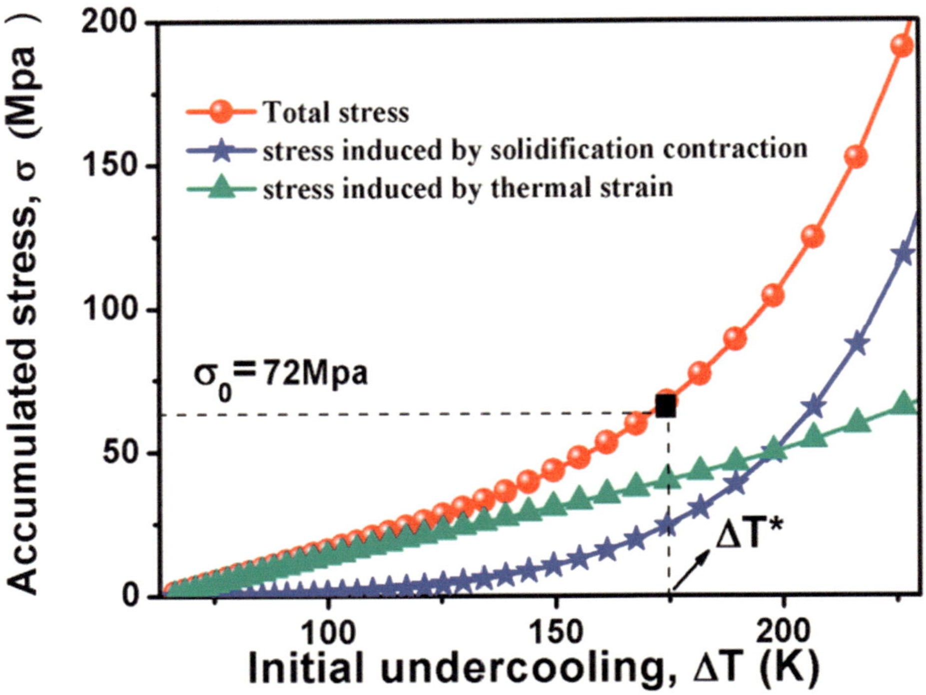

Figure 8. Calculated stress accumulated in dendrite skeleton during rapid solidification of Ni-20at.%Cu alloys as a function of initial undercoolings: the dashed line represents yielding strength the alloys.

microstructures. **Figure 9a–c** show the dense dislocation networks in the grain of Ni-20at.%Cu alloy. The white circle in **Figure 9d** shows an annealing twin in the microstructure. **Figure 8e** shows the electron diffraction spots of the annealing twin in **Figure 9d**. The electron diffraction spots are typical diffraction spot of twins in a FCC alloy. As is known, the stored micro-strain will act as the driving force for recrystallization. Further increasing ΔT causes a continuous increase of the stored micro-strain, resulting in a continuous increase in the extent of deformation. Once the stored energy due to deformation is high enough to initiate the nucleation, the recrystallization will take place. Consequently, a refined recrystallized microstructure can be produced.

Generally, recrystallization is a physical process and consists of two basic processes: the site saturation nucleation which proceeds by atomic thermal activation and the growth of the new strain free grains, when being annealed at temperatures above a proper recrystallization temperature, which is the atomic thermal activation temperature. In order to reveal the recrystallization process in the rapid solidification microstructures of the Ni-20at.%Cu alloys, recrystallization annealing experiments were performed for the quenched and naturally cooled alloys, see in the reference [14]. Compared to the naturally cooled alloys which did not undergo recrystallization. However, as a comparison, the recrystallization drives the microstructures of the quenched alloys to transform into completely newly formed microstructures.

http://dx.doi.org/10.5772/intechopen.70666

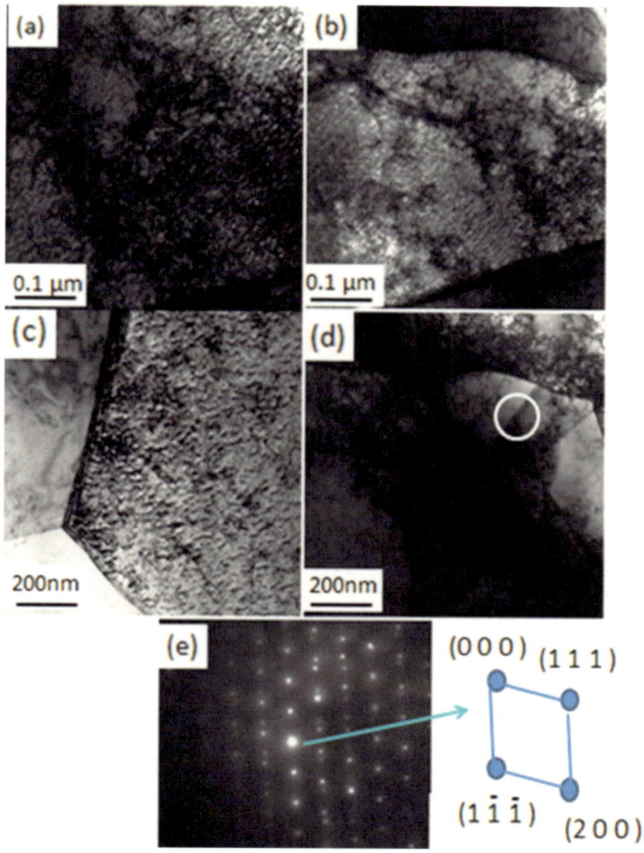

Figure 9. TEM bright field image of (a–d) the dense dislocations and (e) annealing twins in the rapidly solidified $Ni_{80}Cu_{20}$ alloy quenched with a undercooling of about 220 K (220°C).

8. Conclusions

Applying molten glass purification method combined with cyclic superheating method, non-equilibrium microstructural evolution and grain refinement mechanisms of Ni-20at.% Cu and Co-20at.%Pd alloy were investigated systematically. The main conclusions are as follows:

1. Ni-20at.%Cu and Co-20at.%Pd alloy were undercooled by means of fluxing method. Two grain refinement events of the solidification microstructures were observed. The first grain refinement was attributed to chemical superheating induced dendrite remelting. The second grain refinement was ascribed to stress induced recrystallization of the rapidly solidified dendrites.

2. The relationship between the dimensionless superheating and the undercooling indicates that the grain refinement occurring at low undercoolings results from the dendrite break-up

owing to the dendrite remelting, but at high undercooling the dendrite remelting effect is weak.

3. With the help of incorporating the relaxation effect of solute diffusion in bulk undercooled liquid, an extended chemical superheating model for predicting dendrite remelting was developed to explain the two kinds of grain refinement events occurring in the low and high undercooling regimes respectively. It was found that the present dendrite remelting model could predict relatively good results in consistency with the grain refinement event observed in the experiments.
4. It can be inferred that the rapid dendrite growth, the dendrite deformation and the stress-induced dendrite break-up occur during recalescence stage; the solidification of the residual liquid phase, the remelting of dendrites, the recrystallization and the grain growth (secondary recrystallization) occur during post-recalescence period.
5. A corrected model was developed to semiquantitatively calculate the stress accumulation during rapid solidification of undercooled Ni-20at.%Cu alloys. When the undercooling is larger than a critical value, the stress due to solidification contraction and thermal strain will cause break-up of the primary dendrites. The strain energy stored in the broken dendrite pieces will drive recrystallization, leading to a grain refined recrystallization microstructures

Acknowledgements

The authors are grateful to the financial support of National Basic Research Program of China (973Program, 2011CB610403). This work is also supported by International Cooperation Project Supported by Ministry of Science and Technology of China (No. 2011DFA50520).

Conflicts of interest

The authors have no conflicts of interest

Author details

Xiaolong Xu[1,2]*, Hua Hou[1] and Feng Liu[2]

*Address all correspondence to: xiaolong@nuc.edu.cn

1 College of Materials Science and Engineering, North University of China, Taiyuan, Shanxi, People's Republic of China

2 State Key Laboratory of Solidification Processing, Northwestern Polytechnical University, Xi'an, Shaanxi, People's Republic of China

References

[1] Walker JL. The Physical Chemistry of Process Metallurgy, Part 2. New York: Interscience; 1959

[2] Baker JC, Cahn JW. Solidification. 1st ed. Metals Park, OH: ASM; 1971

[3] Dragnevski K, Cochrane RF, Mullis AM. Splitting in deeply undercooled, ultrahigh purity Cu.Physical Review Letters. 2002;**89**:215502

[4] Mullis AM, Cochrane RF. Dendrites Grain refinement and the stability of dendrites growing into undercooled pure metals and alloys. Journal of Applied Physics. 1997;**82**:3783

[5] Powell GLF. Comments on "Undercoolability of copper bulk samples". Journal of Materials Science Letters. 1991;**10**:745

[6] Herlach DM. Materials Science and Engineering R. 1994;**R12**:177-272

[7] Liu F, Yang GC. Stress-induced recrystallization mechanism for grain refinement in highly undercooled superalloy. Journal of Crystal Growth. 2001;**231**:295-305

[8] Karma A. Model of grain refinement in solidfication of undercooled melts. International Journal of Non-Equilibrium Processing. 1998;**11**:201

[9] Li JF, Liu YC, Lu YL, Yang GC, Zhou YH. Structural evolution of undercooled Ni–Cu alloys. Journal of Crystal Growth. 1998;**192**:462-470

[10] Christian JW. The Theory of Transformation in Metals and Alloys. Oxford: Pergamon Press; 2002. p. 832-858

[11] Piccone TJ, Wu Y, Shiohara Y, Flemings MC. Dendritic Growth of Undercooled Nickel-Tin: Part III. Metallurgical Transactions A. 1987;**18A**:925-932

[12] Lu SY, Li JF, Zhou YH. Grain refinement in the solidification of undercooled Ni–Pd alloys. Journal of Crystal Growth. 2007;**309**:103-111

[13] Xu XL, Chen YZ, Liu F. Evidence of recrystallization mechanism of grain refinement in hypercooled Co 80 Pd 20 alloys. Materials Letters. 2012;**81**:73-75

[14] Lu SY, Li JF, Zhou YH. Recrystallization developed in the largely undercooled Ni 54.6 Pd 45.4 alloy. Journal of Alloys and Compounds. 2008;**458**:517-522

[15] Wilde G, Gorler GP, Willnecker R. Specific Heat Capacity of Undercooled Magnetic Melts. Applied Physics Letters. 1996;**69**:2995-2997

[16] Dahle AK, Thevik HJ, Arnberg L, John DH. St. Metallurgical and Materials Transactions B. 1999;**30**:287

[17] Humphreys FJ, Hatherly M. Recrystallization and Related Annealing Phenomena. 2nd ed. New York: Oxford; 2004

[18] Xu XL, Liu F. Observation of recrystallization upon annealing rapidly solidified microstructures of highly undercooled Ni 80 Cu 20 alloys. 2014;**597**:205-210

[19] Galenko PK, Danilov DA. Physics Letters A. Local nonequilibrium effect on rapid dendritic growth in a binary alloy melt. 1997;**235**:271-280

[20] Xu XL, Hou H, Liu F. Nonequilibrium Solidification, Grain Refinements, and Recrystallization of Deeply Undercooled Ni-20 At. Pct Cu Alloys: Effects of Remelting and Stress. Metallurgical and Materials Transactions A: Physical Metallurgy and Materials Science. 2017;1-9

[21] Bhattacharya A, Dutta P. Effect of shrinkage induced flow on binary alloy dendrite growth: An equivalent undercooling model. International Communications in Heat and Mass Transfer. 2014;**57**:216-220

[22] Bhattacharya A, Karagadde S, Dutta P. An equivalent undercooling model to account for flow effect on binary alloy dendrite growth rate. International Communications in Heat and Mass. Transfer. 2013;**47**:15

[23] Thevik HJ, Mo A. The influence of micro-scale solute diffusion and dendrite coarsening upon surface macrosegregation. International Journal of Heat and Mass Transfer. 1997;**40**:2055

[24] Ni J, Beckermann C. A volume-averaged two-phase model for transport phenomena during solidification. Metallurgical Transactions B. 1991;**22**:349

Chapter 3

Coupled Model of Precipitates and Microsegregation During Solidification

Yandong Li, Huamei Duan and Cheng Peng

Additional information is available at the end of the chapter

http://dx.doi.org/10.5772/intechopen.72597

Abstract

To understand the precipitation behavior and solidification process of micro-alloyed steel, abundant thermodynamic data of pure substances were incorporated in the coupled thermodynamic model of inclusions precipitation and solutes micro-segregation during the solidification of heat-resistant steel containing cerium. The liquid inclusions $Ce_{2x}Al_{2y}Si_{1-x-y}O_z$ where $0 < x < 1$, $0 < y < 1-x$ and $z = 1-x-y$ and their generation Gibbs free energy were first introduced to the inclusions reactions according to the Al_2O_3-SiO_2-Ce_2O_3 phase diagram. Then plant trials, lab experiments and published work in the literature were taken account to valid the established model. Also, the difference of calculated results between FactSage and this model were argued. Finally, the liquid inclusions were found in the samples from experiments in the tube furnace based on the calculations by this model.

Keywords: liquid inclusions, thermodynamic model, rare earth, heat-resistant steel

1. Introduction

The heat resistant steels containing rare earth elements (RE) have absolute advantages in mechanical properties and corrosion resistance [1–4]. The oxidation resistance of 253MA steel at high temperature is dramatically enhanced as alloyed with cerium element. However, the rare earth elements are used as deoxidizing and desulfurizing agents in the earlier ages owing to the strong oxidation property [5]. As a result, a large number of inclusions such as Ce_2O_3 and Ce_2O_2S with high melting temperature and small size are generated in the molten steel. During the continuous casting process these inclusions are easily captured by the submerged nozzle. Then the accumulated inclusions with frozen steel lead to nozzle blocking and the production process is interrupted [6–9]. As one of the effectual solutions to nozzle blocking calcium treatment can modify the solid alumina clusters to liquid inclusions which have weak affinity force with the nozzle inner wall [10–13]. Inspired by the above method, the high melting point inclusions containing rare earth elements may be modified to liquid phase by some alloys to solve the nozzle clogging issue of heat resistant steel.

Many prediction models of inclusions formation and evolution are reported in the previous reports. However, the released self-built model mainly involve a few equilibrium reactions of alloy elements with oxygen or sulfur at 1600°C [14, 15]. And the thermodynamic properties of complex liquid inclusions, such as Mn-Si-Al-O, Ca-Al-O, Ca-Al-Mg-O, are calculated by the commercial software [15–19]. Harada assessed the CaO-Al_2O_3 liquid inclusions formation setting the fixed ratio of CaO to Al_2O_3 instead of the whole liquid region [20, 21]. FactSage as the common software in metallurgical field is strong at the equilibrium calculations of multi phases, but the FToxide database is short of solution thermodynamic data containing rare earth and the solidification model is not thoroughly optimized. Liu released one coupling model of micro-segregation and inclusion formation during solidification process in silicon steels [22]. However, the number of reactions was restricted and the solution phase was not considered. Fang proposed a thermodynamic model which includes plenty of formation reactions of rare earth inclusions except the liquid inclusions as a pity [23]. To make up for the deficiency of previous models [24], the more ampler generation reactions of inclusions especially the liquid phase coupling with the solute micro-segregation are introduced in the present model. Then the coupled model was validated by various methods, such as plant trials, lab scale experiments and released experimental data. Finally, the formation regions of liquid phases in the predominance diagrams were given. Moreover, the liquid inclusions with sphere shape and large size were found in the exploitative experiments of lab scale.

2. Modeling process

The coupling model is composed of two parts: the thermodynamic model of inclusions precipitation both in liquid and solid and the dynamic model of solute micro-segregation model in the liquid–solid interface during solidification. The schematic diagram of coupling process is shown in **Figure 1**. Considering the kinetic factors such as diffusion rate and atomic size, the inclusions with large size are formed in the residual liquid phase above the solidus temperature and the small-size carbonitrides are precipitated in the solidified steel under the liquidus temperature. All the relevant reactions of inclusions generation both in liquid and solidified steel are listed in **Table 1** [26–33] and **Table 2** [34–37], respectively. The activity standard state of solute elements and inclusions was 1 mass% infinite dilute solution and the pure substance, respectively. The activity coefficients of solute elements were arranged by Wagner's relations expanded to second order interaction coefficient. The solute segregation at liquid–solid interface during solidification was calculated form the Brody-Flemings micro-segregation model modified by Clyne-Kurz [25]. The more details including the mathematic expressions and thermodynamic properties of solutes can be consulted from the previous works [24].

The generated liquid inclusions whose activity is 1 can be identified as one isolated phase (pure liquid phase) out of the solid inclusions (pure solid phase) and solute elements (solutes in the infinite dilute solution) in the system. As the same as listed in **Table 1**, the liquid inclusion in the form of $Ce_{2x}Al_{2y}Si_{1-x-y}O_z$ (l, $0 < x < 1$, $0 < y < 1\_x$ and $z = 1\_x\_y$) is introduced

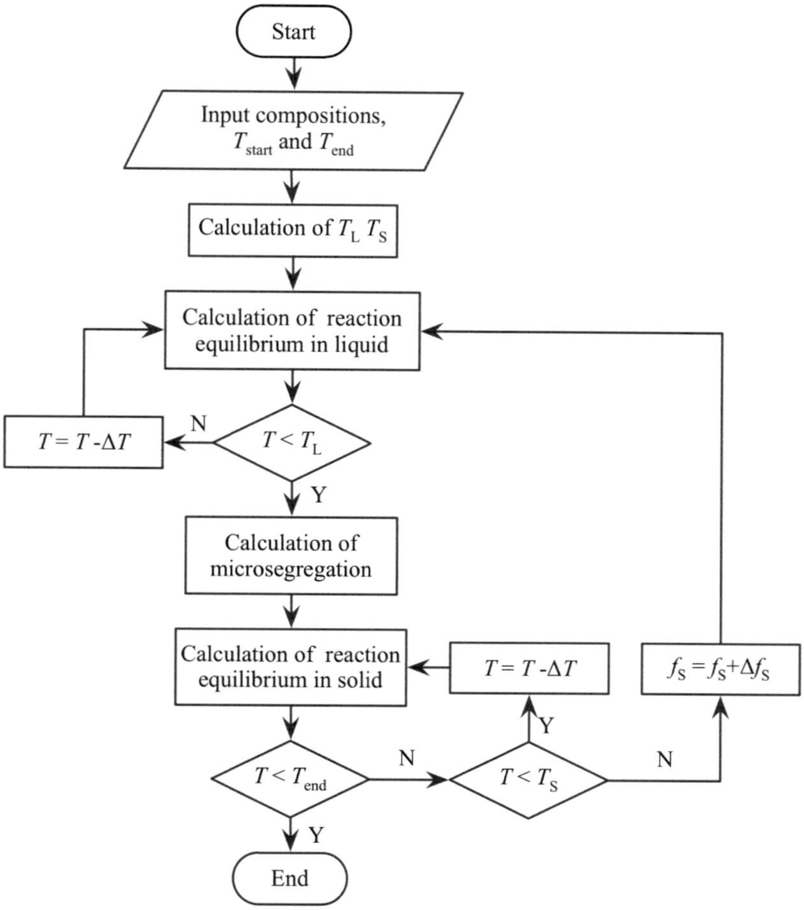

Figure 1. The simplified diagram of coupling process (*fs*—solid fraction).

according to the phase diagram of Al_2O_3-SiO_2-Ce_2O_3 system. The hard task is the calculation of standard Gibbs free energy of $Ce_{2x}Al_{2y}Si_{1\text{-}x\text{-}y}O_z$ generation as the following reaction:

$$x[\mathrm{Ce}] + 2y[\mathrm{Al}] + (1 - x - y)[\mathrm{Si}] + z[\mathrm{O}] = \mathrm{Ce}_{2x}\mathrm{Al}_{2y}\mathrm{Si}_{1-x-y}\mathrm{O}_z(l) \tag{1}$$

The chemical reaction of Eq. 1 can be derived from the following reactions:

$$2[\mathrm{Ce}] + 3[\mathrm{O}] = \mathrm{Ce}_2\mathrm{O}_3(s) \tag{2}$$

$$2[\mathrm{Al}] + 3[\mathrm{O}] = \mathrm{Al}_2\mathrm{O}_3(s) \tag{3}$$

$$[\mathrm{Si}] + 2[\mathrm{O}] = \mathrm{SiO}_2(s) \tag{4}$$

$$\mathrm{Ce}_2\mathrm{O}_3(s) = \mathrm{Ce}_2\mathrm{O}_3(l) \quad \Delta G^{\theta} = 63086.4 - 25.1T \quad J \cdot mol^{-1} \tag{5}$$

$$\mathrm{Al}_2\mathrm{O}_3(s) = \mathrm{Al}_2\mathrm{O}_3(l) \quad \Delta G^{\theta} = 110756.7 - 46.8T \quad J \cdot mol^{-1} \tag{6}$$

Reaction equations in molten steel	$\Delta G^{\theta} = A + B \times T$, J·mol^{-1}	
	A	*B*
$2[Ce] + 3[O] = Ce_2O_3(s)$	−1,431,090	360.06
$[Ce] + 2[O] = CeO_2(s)$	−854274.7	249.11
$[Ce] + [S] = CeS(s)$	−422,783	120.58
$[Ce] + 2[S] = CeS_2(s)$	−131,000	121.4
$2[Ce] + 3[S] = Ce_2S_3(s)$	−1,074,584	328.24
$3[Ce] + 4[S] = Ce_3S_4(s)$	−1,493,010	438.9
$2[Ce] + 2[O] + [S] = Ce_2O_2S(s)$	−1353592.4	331.6
$[Ce] + [Al] + 3[O] = CeAlO_3(s)$	−1,366,460	364
$[Ce] + 11[Al] + 3[O] = CeAl_{11}O_{18}(s)$	−7261120.00	2282.36
$2[Ce] + [Si] + 5[O] = Ce_2SiO_5(s)$	−2077830.00	632.57
$4.67[Ce] + 3[Si] + 13[O] = Ce_{4.67}Si_3O_{13}(s)$	−5299450.00	1641.95
$2[Ce] + 2[Si] + 7[O] = Ce_2Si_2O_7(s)$	−2800700.00	909.66
$2x[Ce] + 2y[Al] + (1-x-y)[Si] + z[O] = Ce_{2x}Al_{2y}Si_{1-x-y}O_z(l)$	—	—
$[Ce] + [N] = CeN(s)$	−401,200	153
$[Ce] + 2[C] = CeC_2(s)$	−202,790	125.3
$2[Ce] + 3[C] = Ce_2C_3(s)$	−224,000	205.8
$[Al] + [N] = AlN(s)$	−267,520	119.54
$2[Al] + 3[O] = Al_2O_3(s)$	−1,205,090	387.72
$[Mn] + [S] = MnS(s)$	−172676.1	56.07
$[Mn] + [O] = MnO(s)$	−284,420	122.75
$[Si] + 2[O] = SiO_2(s)$	−580,550	221.03
$6[Al] + 2[Si] + 13[O] = Al_6Si_2O_{13}(s)$	−4,758,434	1579.22
$4[Cr] + [C] = Cr_4C(s)$	−195,790	218
$23[Cr] + 6[C] = Cr_{23}C_6(s)$	−887889.7	1253.96
$7[Cr] + 3[C] = Cr_7C_3(s)$	−356120.1	417.62
$3[Cr] + 2[C] = Cr_3C_2(s)$	−182030.1	207.39
$2[Cr] + 3[O] = Cr_2O_3(s)$	−797,190	349.72
$3[Cr] + 4[O] = Cr_3O_4(s)$	−94439.9	416.79
$[Cr] + [O] = CrO(s)$	−236,320	113.6
$2[Cr] + [N] = Cr_2N(s)$	−141,300	116.82
$[Cr] + [N] = CrN(s)$	−136,250	96.2
$[Ni] + [O] = NiO(s)$	−125,000	136.29

Table 1. The possible reactions and standard Gibbs free energy in liquid.

Reaction equations in solid phase(γ)	$\Delta G^{\theta} = A + B \times T$, J·mol^{-1}	
	A	*B*
[Al] + [N] = AlN(s)	−137536.1	34.27
[Mn] + [S] = MnS(s)	−176,782	57.91
[Ti] + [N] = TiN(s)	−153158.23	6.13
[Ti] + [C] = TiC(s)	−143585.84	52.65
[Nb] + [N] = NbN(s)	−162730.62	53.6
[Nb] + [C] = NbC(s)	−143585.84	56.67
[V] + [N] = VN(s)	−182030.1	207.39
[V] + [C] = VC(s)	−797,190	349.72

Table 2. The possible reactions and standard Gibbs free energy in solid.

$$SiO_2(s) = SiO_2(l) \quad \Delta G^{\theta} = 9581.3 - 4.8T \quad J \cdot mol^{-1} \tag{7}$$

$$xCe_2O_3(l) + yAl_2O_3(l) + (1 - x - y)SiO_2(l) = Ce_{2x}Al_{2y}Si_{1-x-y}O_z(l) \tag{8}$$

where the standard Gibbs free energy of reactions 2, 3, 4 are listed in **Table 1**. And the CALPHD technique was used to optimize the Gibbs free energy change of reaction 8. The Redlich-Kister (R-K) expression was adopted to describe the excess Gibbs free energy of solution phases in Al_2O_3-SiO_2, Al_2O_3-Ce_2O_3 and Ce_2O_3-SiO_2 binary systems. The R-K expression is written as Eq. 9.

$$\begin{aligned} G^E &= x_1x_2\sum_{j=0}^{n} L_j(x_1 - x_2)^j \\ &= x_1x_2L_0 \\ &+ x_1x_2L_1(x_1 - x_2) \\ &+ x_1x_2L_2(x_1 - x_2)^2 \\ &+ \cdots \end{aligned} \tag{9}$$

where G^E is the excess Gibbs free energy of binary system; x_1, x_2 are the mole fraction of end members in the binary system; L_j (j = 0, 1, 2…) are the interaction parameters of components in the binary system. And L_j in **Table 3** were obtained by the phase diagram optimization when the restricted experimental points from literature were employed.

The excess Gibbs free energy of Ce_2O_3-Al_2O_3-SiO_2 ternary system was derived by the following geometric Kohler method [38].

$L_i = a + bT/(\text{J·mol}^{-1})$		L_0	L_1	L_2	L_3
Al_2O_3-SiO_2	*a*	19570.28	14875.48	5640.02	—
	b	−10.49	−0.71	1.21	—
Ce_2O_3-Al_2O_3	*a*	−78839.95	—	—	—
	b	−30.23	—	—	—
Ce_2O_3-SiO_2	*a*	−140979.96	215301.91	547791.07	398115.65
	b	78.91	23.60	−202.71	−155.94

Table 3. The optimized coefficients of the binary systems.

$$\begin{aligned} G^{E}_{extra} &= (x_1 + x_2)^2 G^E_{12}\left(\frac{x_1}{x_1+x_2}, \frac{x_2}{x_1+x_2}\right) \\ &+(x_2 + x_3)^2 G^E_{23}\left(\frac{x_2}{x_2+x_3}, \frac{x_3}{x_2+x_3}\right) \\ &+(x_1 + x_3)^2 G^E_{13}\left(\frac{x_1}{x_1+x_3}, \frac{x_3}{x_1+x_3}\right) \end{aligned} \tag{10}$$

where the first term $G^E_{12}\left(\frac{x_1}{x_1+x_2}, \frac{x_2}{x_1+x_2}\right)$ is the excess Gibbs free energy at $\left(\frac{x_1}{x_1+x_2}, \frac{x_2}{x_1+x_2}\right)$ point in the 1–2 binary system. The following two terms in Eq. 10 have similar representation as the first term. In order to exactly describe the thermodynamic properties of Ce_2O_3-Al_2O_3-SiO_2 system, the Gibbs free energy was further modified by the $G^E_{ternary}$ term with the value of -825000$x_{Ce_2O_3}$ $x_{Al_2O_3}x^2_{SiO_2}$. After the interaction coefficients in **Table 3** are imported, the standard Gibbs free energy of reaction 8 is written as:

$$\begin{aligned} \Delta G^{\theta}_{L} &= G^{idea} + G^{E} = G^{idea} + G^{E}_{extra} + G^{E}_{ternary} \\ &= -825000xy(1-x-y)^2 + 42569.73xy + 19570.28y(1-y) - 140979.96x(1-x) \\ &+ \frac{y(1-x-y)(1-x-2y)}{1-x}\left(14875.48 + 5640.02\frac{1-x-2y}{1-x}\right) \\ &+ \frac{x(1-x-y)(1-2x-y)}{1-y}\left[215301.91 + 547791.07\frac{1-2x-y}{1-y} + 398115.65\left(\frac{1-2x-y}{1-y}\right)^2\right] \\ &+ \left\{ \begin{aligned} &-98.65xy - 10.49y(1-y) + 78.91x(1-x) \\ &+\frac{y(1-x-y)(1-x-2y)}{1-x}\left[-0.71 + 1.21\frac{1-x-2y}{1-x}\right] \\ &+\frac{x(1-x-y)(1-2x-y)}{1-y}\left[23.6 - 202.71\frac{1-2x-y}{1-y} - 155.94\left(\frac{1-2x-y}{1-y}\right)^2\right] \\ &+8.314[x\ln x + y\ln y + (1-x-y)\ln(1-x-y)] \end{aligned} \right\} T \end{aligned} \tag{11}$$

When the Gibbs free energy of Eqs. (2)–(8) are introduced, the standard Gibbs free energy of Eq. (1) is finally written as follows:

$$
\begin{aligned}
\Delta G_{\mathrm{L}}^{\theta} &= -825000xy(1-x-y)^2 + 42569.73xy + 19570.28y(1-y) - 140979.96x(1-x) \\
&- 1368003.6x - 1094333.3y - 570968.7(1-x-y) \\
&+ \frac{y(1-x-y)(1-x-2y)}{1-x}\left(14875.48 + 5640.02\frac{1-x-2y}{1-x}\right) \\
&+ \frac{x(1-x-y)(1-2x-y)}{1-y}\left[215301.91 + 547791.07\frac{1-2x-y}{1-y} + 398115.65\left(\frac{1-2x-y}{1-y}\right)^2\right] \\
&+ \left\{\begin{aligned}
&-98.65xy - 10.49y(1-y) + 78.91x(1-x) + 334.96x + 340.92y + 216.23(1-x-y) \\
&+ \frac{y(1-x-y)(1-x-2y)}{1-x}\left[-0.71 + 1.21\frac{1-x-2y}{1-x}\right] \\
&+ \frac{x(1-x-y)(1-2x-y)}{1-y}\left[23.6 - 202.71\frac{1-2x-y}{1-y} - 155.94\left(\frac{1-2x-y}{1-y}\right)^2\right] \\
&+8.314[x\ln x + y\ln y + (1-x-y)\ln(1-x-y)]
\end{aligned}\right\} T
\end{aligned} \tag{12}
$$

After the standard Gibbs free energy $\Delta G_{\mathrm{L}}^{\theta}$ of liquid inclusions formation was worked out, the inclusions compositions (certain x, y values, $z = 1-x-y$) can be calculated according to the minimum of Gibbs free energy ΔG_{L}. At the equilibrium status in the molten steel, the Gibbs free energies of chemical reactions in **Table 1** are greater than or equal to zero.

3. Modeling validation

The constructed model was validated by plant trial, lab scale experiments and experimental data from others work. Moreover, the calculated results by this model were compared with the popular commercial software FactSage.

The standard chemical compositions of 253MA steel were shown in **Table 4**. And the industrial samples of 253MA steel were obtained after EAF → AOD → LF → CC process in one stainless steel plant of China. The obtained samples were machined into column of $\Phi = 4$ mm

C	Si	Mn	S	P	N	Ni	Cr	Ce
0.05–0.10	1.40–2.0	≤0.80	≤0.030	≤0.040	0.14–0.20	10.00–12.00	20.00–22.00	0.03–0.08

Table 4. The chemical compositions of 253MA steel (mass fraction, %).

and then polished by emery paper and polishing cloth to make metallographic specimen. Then the field emission scanning electron microscope (FE-SEM) and energy dispersive spectrometer (EDS) were employed to observe the morphology and analyze the compositions of inclusions, as shown in **Figure 2**. It can be seen that the inclusions with the size of 1-2 μm in the solidification structure of 253 MA steel are mainly Ce_2O_2S and Ce_2O_3. CeN maybe exist as the element N was detected as shown in **Figure 2(b)**. The peaks without labels are matrix elements Fe, Cr and Ni.

The experiments in laboratory that carefully follows the industrial operations were carried out in the Si-Mo rod heating furnace protected by argon flow at 1873 K, as shown in **Figure 3**. Firstly, the Fe-Cr-Ni alloys with fixed ratio as in the 253MA steel were loaded in the MgO crucible until completely melted. Then the mass fraction of Si was calculated and added into the crucible after the oxygen content was first determined. After Si was totally melted, the oxygen content was determined for the second time and then the melt was alloyed by Ce. The molten steel was drawn using quartz tube (inner $\Phi = 4$ mm) and immediately quenched into ice water at 30 min after Ce addition. Therefore, the inclusions formed at 1873 K can be preserved since the extremely cooling rate of quenching. The quenched samples were processed and observed as the same as that of industrial samples. It is explicit that the inclusions are Ce_2O_2S and Ce_2O_3 with the size smaller than 2 μm as the analysis results shown in **Figure 4**.

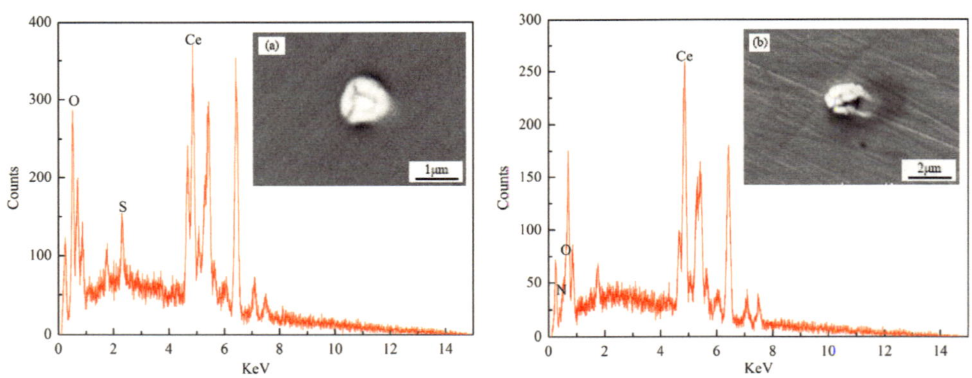

Figure 2. SEM images and EDS analysis of inclusions in the solidification structure.

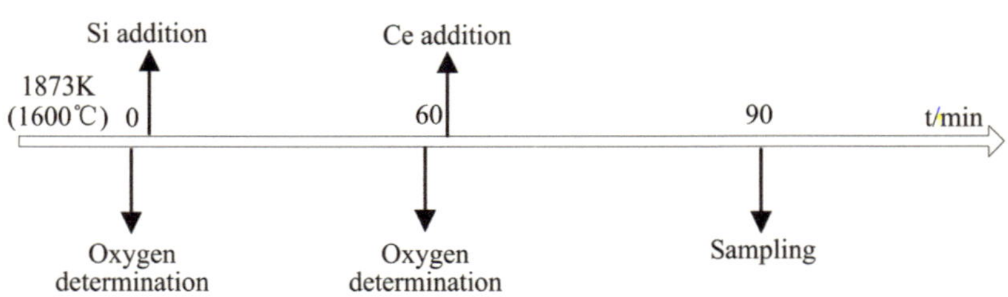

Figure 3. The flow chart of technological process of lab scale experiment.

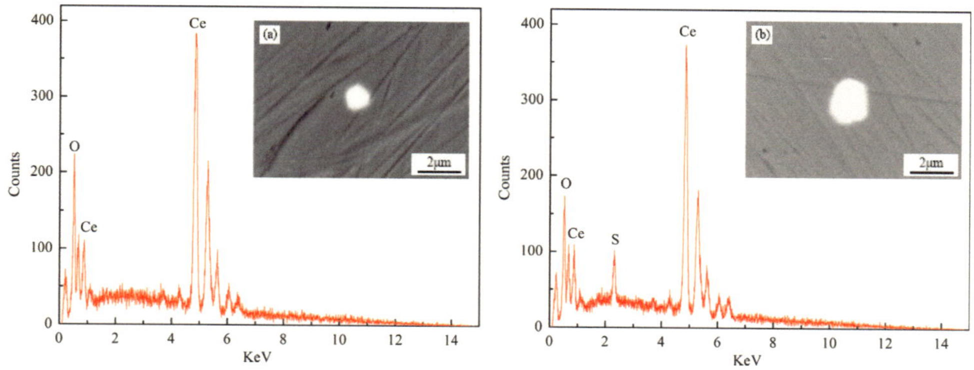

Figure 4. SEM images and EDS analysis of inclusions in the laboratory sample.

The types and amounts of inclusions formed during the solidification of 253MA steel was calculated by the present model and the results are drawn in **Figure 5** which (b) is the zoom view of the liquid–solid two-phase region ($T_L = 1713$ K). It can be seen that the inclusions types are mainly Ce_2O_2S, Ce_2O_3, CeN and Ce_2SiO_5. And the calculation results are in agreement with the industrial experiments except Ce_2SiO_5. Ce_2SiO_5 is formed as Ce_2O_3 is reduced during solidification as shown in the magnified image of **Figure 5(b)**. Then Ce_2SiO_5 must be transformed from Ce_2O_3 as the increasing Si concentration in the residual liquid phase which owes to the micro-segregation of solutes at the liquid–solid interface. However, it is very difficult for the decomposition of Ce_2O_3 because the extremely fast cooling rate during the continuous casting process. Therefore, Ce_2SiO_5 is not found in the industrial samples. What is more, the formed inclusions are only Ce_2O_3 and Ce_2O_2S at 1873 K which is completely consistent with the results in the laboratory scale experiments.

The types and amounts of inclusions formed during the solidification of 253MA steel was also calculated by FactSage and the results are drawn in **Figure 6**. It can be seen that only Ce_2O_3 and Ce_2O_2S are formed in the FactSage results where Ce_2O_2S precipitates at a lower temperature 1848 K comparing with 1873 K in the present model. However, Ce_2SiO_5 and CeN are not present in the FactSage results since the solutes segregation is not considered in FactSage when

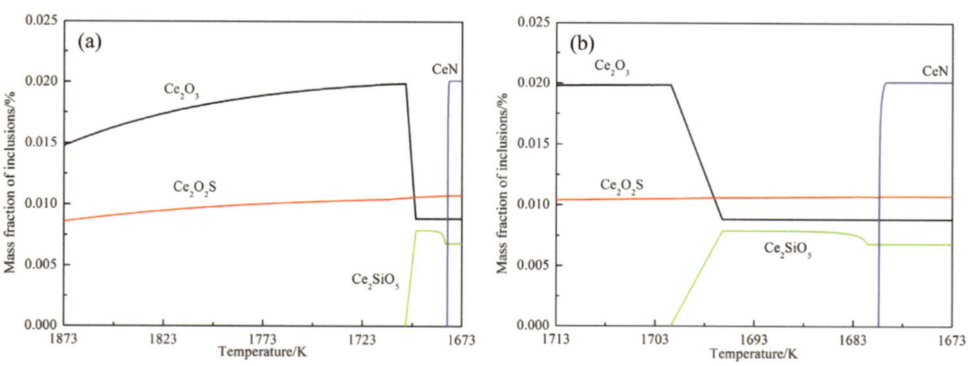

Figure 5. The precipitation behavior of inclusions during solidification of 253MA steel.

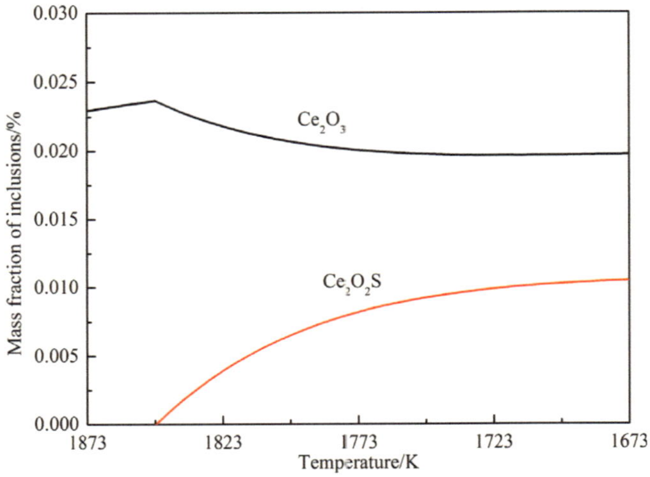

Figure 6. The precipitation behavior of inclusions calculated by FactSage.

the normal equilibrium solidification model is selected. The elements activities were further calculated to understand the lower formation temperature of Ce_2O_2S in FactSage. The stable pure substances at standard conditions are adopted by FactSage as the activities standard sate when 1mass% infinite dilute solution is chosen in the present work. The activities of solutes in molten steel calculated by the two different standard states can be linked by the following equation.

$$a_{i(w)}^{\mathrm{H}} = \frac{100M_i}{M_{\mathrm{A}}\gamma_i^0} * a_i^{\mathrm{R}} \tag{13}$$

where $a_{i(w)}^{\mathrm{H}}$ is the activity at the standard state of 1 mass% solution; a_i^{R} is the activity when pure substance is standard state; M_i is the relative atomic mass of solutes; M_{A} is the relative atomic mass of solvents, that is Fe in this work; γ_i^0 is the activity coefficient when stable pure substance is standard state in the Henrian solution. The available activity coefficients are listed in **Table 5** [39] and the activities transformed from FactSage state to 1 mass% solution comparing with that calculated by this model are present in **Table 6**.

As listed in **Table 6**, the solutes activities calculated from different standard state are basically at the same level except P and N elements. The remarkable differences of P and N activities can

	C	Si	Mn	Ce	O
A	$e^{0.57}$	−11763.2	0	−7103.2	−15,280
B		0		0	3.5
Standard state	S	S	L	L	G

Note: $\ln\gamma^0 = A/T + B$, *e*: natural constant, S: pure solid, L: pure liquid, G: gas.

Table 5. Optimized Henrian activity coefficients $\ln\gamma^0$.

	C	Si	Mn	P	S	N	Ce	O
This work	4.23E-2	3.10	5.74E-1	2.78E-2	7.15E-4	1.50E-2	2.09E-3	2.82E-4
FactSage	3.08E-2	2.64	5.20E-1	9.30E-5	1.39E-4	13.2	1.24E-3	1.86E-4

Table 6. The activities calculated by this work and FactSage at 1873 K.

Authors	RE addition/%	Inclusions found	Predicted by this model
Yue [40]	0.016	Rare earth sulfides	Ce_2S_3
Liu [41]	0.044	Rare earth oxysulfides	Ce_2O_2S, CeS
Guo [42]	0.034	Ce_2O_2S, CeS	Ce_2O_2S, Ce_2O_3, CeN
Zhai [43]	0.08	Ce_2O_2S, Ce_2O_3, CeO_2	Ce_2O_2S, Ce_2O_3
Factsage	0.04	Ce_2O_3, Ce_2O_2S	Ce_2O_3, Ce_2O_2S, Ce_2SiO_5, CeN
Present work	0.04	Ce_2O_3, Ce_2O_2S, CeN	Ce_2O_3, Ce_2O_2S, Ce_2SiO_5, CeN

Table 7. Comparison of inclusions from literature experiments and predicted by this work.

owe to the missing γ_i^0 values of P and N. It can be seen that the sulfur activity calculated by FactSage is smaller than that calculated by this model because γ_i^0 of S is not available. It is the direct explanation why the formation temperature of Ce_2O_2S is lower in the FactSage results.

Besides the industrial trials and lab scale experiments, the types of inclusions reported in the steel containing rare earth elements from literature were compared with the model prediction and the results are present in **Table 7**. The types of generated inclusions in various grades of steels reported in others work are almost the same as the results calculated by this model.

All the above comparisons send the message that the accomplished coupled thermodynamic in this work is precise and it is competent to exactly assess the precipitation behaviors of inclusions during solidification process of various grades of steels containing rare earth elements.

4. Model application

Aluminum was alloyed in the heat-resistant steel imitating the calcium treatment way to solve the nozzle clogging of 253MA steel. Since the formed inclusions resulting in nozzle clogging are mainly oxides, the inclusions transformation at different initial oxygen contents was studied. And the types of inclusions formed before and after aluminum addition are shown in **Figure 7**. It can be seen that the stable inclusion at high oxygen content is $Ce_2Si_2O_7$ in steel without aluminum. By contrast, the inclusions eventually transform to liquid phase as oxygen content increases to 0.018% after 0.01mass% aluminum was added in 253MA steel.

In addition, the stable areas of inclusions (predominance area diagram of inclusions or phase diagram of inclusions) in heat-resistant steel at 1873 K are summarized in **Figure 8**. The stable

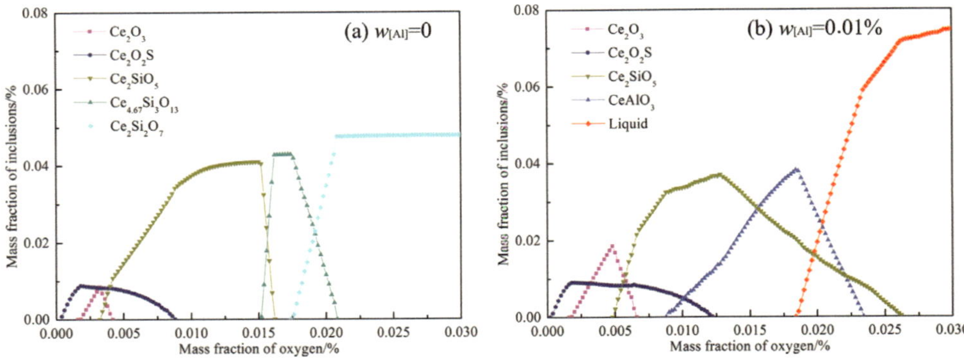

Figure 7. The inclusions types in heat-resistant steel at 1873 K.

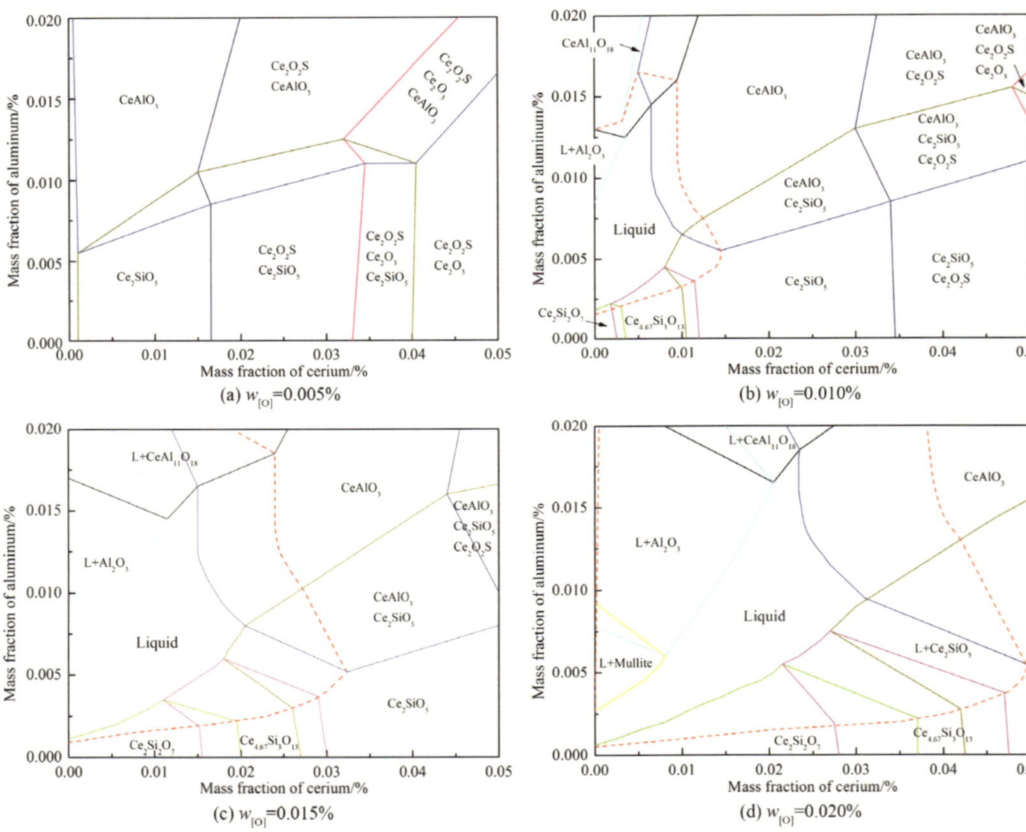

Figure 8. The stability areas of inclusions at 1873 K.

regions of liquid inclusions are surrounded by the dash line. There is not liquid inclusions formed when the oxygen content is low (0.005% in this study). And the stable areas of liquid inclusions are growing as the increasing initial oxygen content. There can be no doubt that

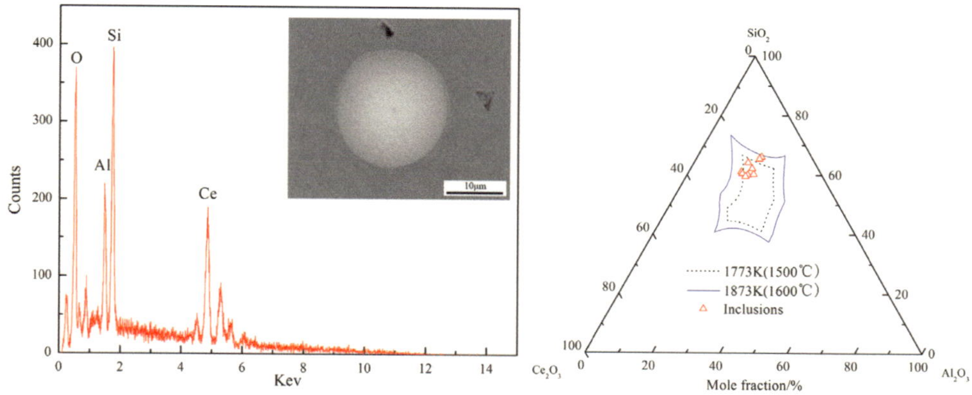

(a) SEM image and EDS analysis
(b) Compsition of found inclusions

Figure 9. The liquid inclusions in heat-resistant.

stable areas of liquid phase will be helpful to solve the nozzle clogging issue of steels alloyed by rare earth elements.

In 2011, Kojola [6] reported that the clogging rate of 253MA steel dramatically decreased when Al, Si and Ce co-existed in the molten steel. Kojola argued that the small rare earth oxides can be dissolved after Si addition because of the negative interaction of Si to oxygen. However, no convincing proof was given. It is clear that the liquid inclusions are formed according to the prediction by the model in this work. Moreover, based on the calculated stable areas of inclusions the lab scale experiments following the procedure in **Figure 3** were carried out when Si-Al alloy was chosen instead of pure Si. The FE-SEM image and EDS analysis of found inclusions were shown in **Figure 9(a)**. The inclusions were complex phases of Ce-Al-Si-O system with the large size of 15 μm. And the complex inclusions were in the liquid region at 1873 K when be marked in the Al_2O_3-SiO_2-Ce_2O_3 phase diagram.

5. Conclusions

A coupling model of inclusions generation and solutes micro-segregation during the solidification process of steels containing rare earth elements was worked out. The method for calculating the Gibbs free energy of liquid inclusions was first given. Then the reliability of mathematical model was validated by the industrial sampling, experiments in lab scale and inclusions found in literature. The inclusions calculated by FactSage software were different from this model because of the differences in the activity standard state and the solidification model. The inclusions are Ce_2O_3, Ce_2O_2S and CeN in the solidification structure of 253MA steel. These small size inclusions containing rare earth elements can be modified to the liquid inclusions after Al adding and the stable liquid regions increase as the initial oxygen content increases. Moreover, the complex liquid inclusions of Ce-Al-Si-O system were found in the preliminary experiments of lab scale. The coupling model is helpful to relieve the nozzle clogging issue of various grades of steel containing rare earth elements.

Acknowledgements

This work was supported by the Scientific Research Foundation for Scholars in Yangtze Normal University No. 2016KYQD14, No. 2016XJQN34.

Author details

Yandong Li[1]*, Huamei Duan[2] and Cheng Peng[1]

*Address all correspondence to: andyydlee@gmail.com

1 Key Laboratory of Extraordinary Bond Engineering and Materials, Yangtze Normal University, Chongqing, China

2 College of Materials Science and Engineering, Chongqing University, Chongqing, China

References

[1] Chen L, Ma XC, Wang LM, Ye XN. Effect of rare earth element yttrium addition on microstructures and properties of a 21Cr-11Ni austenitic heat-resistant stainless steel. Materials and Design. 2011;**32**(4):2206-2212. DOI: 10.1016/j.matdes.2010.11.022

[2] Matway RJ, McGuire MF, Mehta J. Steel alloy having improved creep strength. US. Patent. 1995

[3] Nilsson JO. The influence of nitrogen on high temperature low cycle fatigue behaviour of austenitic stainless steels. Fatigue & Fracture of Engineering Materials & Structures. 1987;**7**(1):55-64. DOI: 10.1111/j.1460-2695.1984.tb00359.x

[4] Cai GJ, Li CS. Effects of Ce on inclusions and corrosion resistance of low-nickel austenite stainless steel. Materials and Corrosion. 2015;**66**(12):1445-1455. DOI: 10.1002/maco.20150 8380

[5] Waudby PE. Rare earth additions to steel. International Metals Reviews. 1978;**23**(1):74-98. DOI: 10.1179/imtr.1978.23.1.74

[6] Kojola N, Ekerot S, Andersson M, Jönsson P. Pilot plant study of nozzle clogging mechanisms during casting of REM treated stainless steels. Ironmaking and Steelmaking. 2011;**38**(1):1-11. DOI: 10.1179/030192310X12690127076398

[7] Kojola N, Ekerot S. Pilot plant study of clogging rates in low carbon and stainless steel grades. Ironmaking & Steelmaking. 2011;**38**(2):81-89. DOI: 10.1179/030192310X12706364542704

[8] Roos E, Karasev A, Jönsson PG. Effect of Si and Ce contents on the nozzle clogging in a REM alloyed stainless steel. Steel Research International. 2015;**86**(11):1279-1288. DOI: 10.1002/srin.201400344

[9] Yao YK, Zhu MW, Wang DY, Liu CJ, Jiang MF. Investigation of nozzle blockage mechanism during feeding RE into tundish. Chinese Eare Earth. 2004;**25**(5):17-19. DOI: 10.16533/j.cnki.15-1099/tf.2004.05.006

[10] Lind M, Holappa L. Transformation of alumina inclusions by calcium treatment. Metallurgical and Materials Transactions B. 2010;**41**(2):359-366. DOI: 10.1007/s11663-009-9337-9

[11] Janke D, Ma Z, Valentin P, Heinen A. Improvement of castability and quality of continuously cast steel. ISIJ International. 2000;**40**(1):31-39. DOI: 10.2355/isijinternational.40.31

[12] Holappa L, Hämäläinen M, Liukkonen M, Lind M. Thermodynamic examination of inclusion modification and precipitation from calcium treatment to solidified steel. Ironmaking & Steelmaking. 2003;**30**(2):111-115. DOI: 10.1179/030192303225001748

[13] Yang J, Wang XH, Jiang M, Wang WJ. Effect of calcium treatment on non-metallic inclusions in ultra-low oxygen steel refined by high basicity high Al_2O_3 slag. Journal of Iron and Steel Research, International. 2011;**18**(7):8-14. DOI: 10.1016/S1006-706X(11)60083-6

[14] Zhang LF, Ren Y, Duan HJ, Yang W, Sun LY. Stability diagram of Mg-Al-O system inclusions in molten steel. Metallurgical and Materials Transactions B. 2015;**46**(4):1809-1825. DOI: 10.1007/s11663-015-0361-7

[15] Zhang TS, Min Y, Jiang MF. Effect of magnesium addition on evolution of inclusions in Mn–Si–Al deoxidised molten steels. Canadian Metallurgical Quarterly. 2015;**54**(2):161-169. DOI: 10.1179/1879139514Y.0000000173

[16] Fujii K, Nagasaka T, Hino M. Activities of the constituents in spinel solid solution and free energies of formation of MgO, $MgO{\cdot}Al_2O_3$. ISIJ International. 2000;**40**(11):1059-1066. DOI: 10.2355/isijinternational.40.1059

[17] Park JH, Todoroki H. Control of MgO Al_2O_3 spinel inclusions in stainless steels. ISIJ International. 2010;**50**(10):1333-1346. DOI: 10.2355/isijinternational.50.1333

[18] Ren Y, Zhang LF, Yang W, Duan HJ. Formation and thermodynamics of Mg-Al-Ti-O complex inclusions in Mg-Al-Ti-deoxidized steel. Metallurgical and Materials Transactions B. 2014;**45**(6):2057-2071. DOI: 10.1007/s11663-014-0121-0

[19] Zhang TS, Min Y, Liu CJ, Jiang MF. Effect of Mg addition on the evolution of inclusions in Al–Ca deoxidized melts. ISIJ International. 2015;**55**(8):1541-1548. DOI: 10.2355/isijinternational.ISIJINT-2014-691

[20] Harada A, Maruoka N, Shibata H, Kitamura S. A kinetic model to predict the compositions of metal, slag and inclusions during ladle refining: Part 1. Basic concept and application. ISIJ International. 2013;**53**(12):2110-2117. DOI: 10.2355/isijinternational.53.2110

[21] Harada A, Maruoka N, Shibata H, Kitamura S. A kinetic model to predict the compositions of metal, slag and inclusions during ladle refining: Part2. Condition to control the inclusion composition. ISIJ International. 2013;**53**(12):2118-2125. DOI: 10.2355/isijinternational.53.2118

[22] Liu ZZ, Wei J, Cai KK. A coupled mathematical model of microsegregation and inclusion precipitation during solidification of silicon steel. ISIJ International. 2002;**42**(9):958-963. DOI: 10.2355/isijinternational.42.958

[23] Fang L, Liu HL, Liu CJ, Jiang MF. Model analysis on behavior of cerium in heavy rail steel. Chinese Rare Earth. 2008;**29**(1):88-91. DOI: 10.3969/j.issn.1004-0277.2008.01.022

[24] Li YD, Liu CJ, Li CL, Jiang MF. A coupled thermodynamic model for prediction of inclusions precipitation during solidification of heat-resistant steel containing cerium. Journal of Iron and Steel Research, International. 2015;**22**(6):457-463. DOI: 10.1016/S1006-706X(15)30027-3

[25] Clyne TW, Kurz W. Solute redistribution during solidification with rapid solid state diffusion. Metallurgical and Materials Transactions A. 1981;**12**(6):965-971. DOI: 10.1007/BF02643477

[26] Vahed A, Kay DAR. Thermodynamics of rare earths in steelmaking. Metallurgical Transactions B. 1976;**7**(3):375-383. DOI: 10.1007/BF02652708

[27] Turkdogan ET. Physical Chemistry of High Temperature Technology. New York: Academic Press; 1980

[28] Alaoua D, Lartigue S, Larere A, Priester L. Precipitation and surface segregation in low carbon steels. Materials Science and Engineering A. 1994;**189**(1–2):155-163. DOI: 10.1016/0921-5093(94)90411-1

[29] Yamamoto K, Hasegawa T, Takamura JI. Effect of boron on intra-granular ferrite formation in Ti-oxide bearing steels. ISIJ International. 1996;**36**(1):80-86. DOI: 10.2355/isijinternational.36.80

[30] Taylor KA. Solubility products for titanium-, vanadium-, and niobium-carbide in ferrite. Scripta Metallurgica et Materialia. 1995;**32**(1):7-12

[31] ParK JY, Park JK, Choo WY. Effect of Ti addition on the potency of MnS for ferrite nucleation in C–Mn–V steels. ISIJ International. 2000;**40**(12):1253-1259. DOI: 10.2355/isijinternational.40.1253

[32] Chen EB, Dong YC, Guo SZ. Study on thermodynamical properties in Mn-Fe alloy melts. Acta Metallgugica Sinica. 1997;**33**(8):831-837

[33] Du T. Themodynamics of rare earth elements in iron-base solutions. Journal of Iron and Steel Research. 1994;**6**(3):6-12. DOI: 10.13228/j.boyuan.issn1001-0963.1994.03.002

[34] Yue EB, Qiu ST, Gan Y. Carbo-nitride precipitation model in austenite of HSLA steel. Journal of Iron and Steel Research. 2006;**18**(8):49-52. DOI: 10.3321/j.issn:1001-0963.2006.08.012

[35] Tang GB, Wu XY, Yong X, Bai AM, Liu ZD, Yong QL. Numerical simulation of thermodynamic for carbonitride precipitationin austenite of complex microalloyed high strength and low alloy steels. Heat Treatment Metals. 2008;**33**(8):67-72. DOI: 10.13251/j.issn.0254-6051.2008.08.044

[36] Luo S, Zhu MY, Ji C, Cai ZZ. Solute microsegregation model for continunous casting process of steel. Iron and Steel. 2010;**45**(6):31-36

[37] Zhang MC, Li SZ, Zhang QY, Zhang JP. Thermodynamic study on carbide and nitride precipitation of X60 pipeline steel in continuous casting. Journal of Anhui University of Technology (Natural Science). 2012;**29**(3):207-210. DOI: 10.3969/j.issn.1671-7872.2012.03.004

[38] Pelton AD. A general "geometric" thermodynamic model for multicomponent solutions. Calphad. 2001;**25**(2):319-328. DOI: 10.1016/S0364-5916(01)00052-9

[39] Jung IH, Decterov SA, Pelton AD. A thermodynamic model for deoxidation equilibria in steel. Metallurgical and Materials Transactions B. 2004;**35**(3):493-507. DOI: 10.1007/s11663-004-0050-4

[40] Yue LJ, Wang LM, Han JS. Effects of rare earth on inclusions and corrosion resistance of 10PCuRE weathering steel. Journal of Rare Earths. 2010;**28**(6):952-956. DOI: 10.1016/S1002-0721(09)60219-2

[41] Liu X, Yang JC, Gao XZ. Effects of RE on the inclusions and impact toughness of 2Cr13 stainless steel. Journal of University of Science and Technology Beijing. 2010;**32**(5):605-609

[42] Guo F, Lin Q, Sun XX. Thermodynamic calculation on precipitation and transformation of inclusions in high purity carbon-manganese steels with rare earths and their microstructure analysis. Journal of the Chinese Rare Earth Society. 2004;**22**(5):624-618. DOI: 10.3321/j.issn:1000-4343.2004.05.006

[43] Zhai QJ. Study on microstructure and formation mechanism of rare earth inclusions during solidification (Chinese edition). Journal of the Chinese Rare Earth Society. 1995;**13**(1):60-62

Chapter 4

Solidification Physics and Microstructure: A Study of AlMg and AlMgSi Alloys by Vortex Method

Socorro Valdez Rodríguez, Lorenzo Martínez and
Martín Pech-Canul

Additional information is available at the end of the chapter

http://dx.doi.org/10.5772/intechopen.72763

Abstract

Physical properties on solidification process deal with the internal atomic arrangement, named generally microstructure. The microstructure characteristic is essential for the properties of metallic materials, including binary, ternary or eutectic alloys. Thus, despite significant progress on microstructural evolution, numerous challenges still exist for revealing the internal structure to act synergistically with matrix alloy chemical components. It is well known that the internal microstructure of an alloy controls and modifies the nature, interaction type and properties of the existing defects and final applications. At this point, necessary understanding of the correlation between equilibrium and non-equilibrium effects, surface energy and chemical potential for the time of the structures formation is probably the key to solidification that can help to predict the complex of microstructures. For the case of multi-phase solidification, AlMgSi alloy involves thousands of atoms, at the atomic scale, that transit to microstructures while the solid-liquid phase transformation occurs. This chapter outlines the role of solidification physics, and the atomic arrangement is believed to have in influencing mechanical hardness properties and degradation resistance.

Keywords: molecular interaction, microstructure, mechanical properties, mass transport, cooling rate

1. Introduction

The purpose of this chapter is to underlie the atomistic theory of AlMg alloy solidification, emphasizing the aspects that relate to microstructure arrangement with the addition of silicon as microalloying. The microstructure characteristic is essential for the properties of metallic materials, including binary, ternary or eutectic alloys. This is a problem of pronounced real-world significance as structural materials and also touches on some of the deepest issues of physics like the atomic transport, expressed here in terms that are accessible to a beginner with no background in quantum mechanics. Even though the microstructure considered as a different atomic combination, it reduces to a mass equation [1]. With this in mind, the chapter treaties with the solidification physics, that related the maturity of physically based models of the understanding of AlMg and AlMgSi microstructures. The effect of magnesium as microalloying element has a pronounced solid solubility in aluminum, which develops a high precipitation-hardening capability [2]. The outstanding attributes to AlMg base alloy can increase the motivation to substitute some ferrous alloys with lightweight aluminum-magnesium alloys. Some delicately balanced between experimental and the atomistic theory resulting in an alternative to develop technologies more competitive as production process in order to reduce manufacturing costs and improve material performance.

To understand the importance of aluminum and aluminum alloys, their valuable engineering purposes in automotive parts and aircraft components associated with their lightweight should be mentioned [3–6]. In addition, the aluminum and their alloys possess a good performance, castability, high mechanical strength properties and attractive resistance to electrochemical surface degradation [7–9]. Many investigators describe these advantages as primary kind of light alloys since structural design, architectural applications, to combustion engines, until tribological functions [10–13]. Recently, AlMg alloys have been applied as matrix alloy to composite materials due to their low density and high wettability [14, 15]. Therefore, the development of AlMg alloys is an important task, and the differences between binary, ternary and quaternary alloy make them an important research topic. Interest in the addition of silicon, iron, manganese and silver to the aluminum magnesium matrix is motivated by the marked effect on their corrosion resistance, mechanical behavior and atomic structure arrangement [16, 17]. Thus, despite significant progress on microstructural evolution, numerous challenges still exist for revealing the internal structure to act synergistically with matrix alloy chemical components. Moreover, the alloying elements, such as silicon, exhibited unique structural arrangement that elevated the mechanical behavior reached by precipitation development occurred after thermal treatment [18]. The precipitation in these alloys indicates a several transformation sequences from the supersaturated solid solution that relies on the fine precipitation, which strained the crystallographic planes of the Al-matrix and improving the specific Young's modulus, high specific strength, yield and creep strengths [19–22]. The silicon and iron microalloying can refine microstructure of AlMg alloys due to the formation of Mg_2Si and Al_2MgC_2 phases during the solidification process [23, 24]. Interest in the addition of silicon and Sc to the aluminum magnesium matrix is motivated by the marked effect on their corrosion resistance, mechanical behavior, and atomic structure arrangement [25, 26]. Moreover, various averaging procedures suggest that AlMgSi alloys reach up to 90% extruded volume owing to an attractive combination of mechanical and chemical properties and an excellent response to surface finishing operations [27–29]. A gradual increase of extruded pieces continued growth by the 6000 series like in automotive, marine and architectural applications [13, 21].

Thus, despite significant progress on microstructural evolution, numerous challenges still exist for revealing the internal structure to act synergistically with matrix alloy chemical components [3]. It is well known that the internal microstructure of an alloy controls and modifies the nature, interaction type and properties of the existing defects and final applications [4, 25].

In general, the physical properties on solidification process deal with the evolution of microstructures during the solid-liquid phase transformation. Additionally, materials science involves the solid-liquid transformation in the manufacturing of metallic alloys, quasicrystal synthesis, composite materials and the new entropic alloy systems [16]. Furthermore, the solidification process has been a key factor in modern technologies for obtaining casting parts that have been based on physical and mechanical treatment applied to alloys in a molten condition [27, 28]. Therefore, the velocity solidification, related with solute partition, influences the solid-liquid interface atomic diffusion [9]. The internal atomic arrangement, named generally microstructure, as retrieved from the mold, is significant for new materials design and new manufacturing process. It seems that with the appropriate chemical alloying, a relevant material properties could lead to microstructural influences in the various properties including friction and wear, thermal, mechanical, electrical, electrochemical degradation, corrosion resistance, formability and welding characteristics. This chapter outlines the role of solidification physics, and the atomic arrangement is believed to have in influencing mechanical hardness properties and degradation resistance.

2. Atomistic view of solidification of AlMgSi alloy

One of its main supposed advantages is the conceptual framework, which underlies the atomistic or microscopic theory of matter that conveys of atoms building micro and nanostructures [14]. In keeping with the research, continues a few practical examples to design engineer materials and devices on an atomic scale as well as to predict the structure of new materials; the flow of current through a nanostructure; their electrical, physical and mechanical properties and the rates of chemical reactions [11–13]. Here, we would like to highlight the microstructural arrangement that makes major contributions in the condensed matter to the development of crystalline and electronic structures correlated with wear life, resistance to reactive degradation, novel electronic devices and nonequilibrium entropic alloys [4, 15]. At this point, necessary understanding of the correlation between equilibrium and non-equilibrium effects, surface energy and chemical potential for the time of the structures formation is probably the key to solidification that can help to predict the complex of microstructures.

The multi-phase solidification of AlMgSi alloy or melting this material for small-scale physical phenomena to incorporate in large-scale models implies a multi-physics nature of solidification [26]. The solidification process involves hundreds of thousands of atoms, related with their melting point, when nucleation initiates from solid-like regions, of the new phase, on the atomic scale until nuclei growths until a crystal during a large fraction of microsecond. In the formation of the solid phase, the surface energy contributes to instability of the liquid/solid interface [27]. At atomic scale, the conjunction of crystals proceeds in a complementary network of grains with heterogeneities like voids, vacancies, dislocations and evidently grain boundaries.

2.1. AlMgSi phases

Taking into account our microstructural data material, firstly the AlMg alloy with the phase diagram from **Figure 1**, their essential phase corresponds to the α and β solid phases of Al and Mg, respectively, and the presence of intermetallic phases. One can observe regions where

only liquid phase l exists,

only solid phase α exists,

liquid l and solid phase α coexist,

liquid l and solid phase β coexist,

both solid phases α and β coexist.

Taking this last one consideration into account, the growth in that region is also known as eutectic solidification. As we can observe in **Figure 2**, for AlMgSi alloys phase diagram, intermetallics and grain boundary phases as consequence of thermal treatment process are identified. As one of the principal, intermetallic appears with iron-containing intermetallic and MgSi

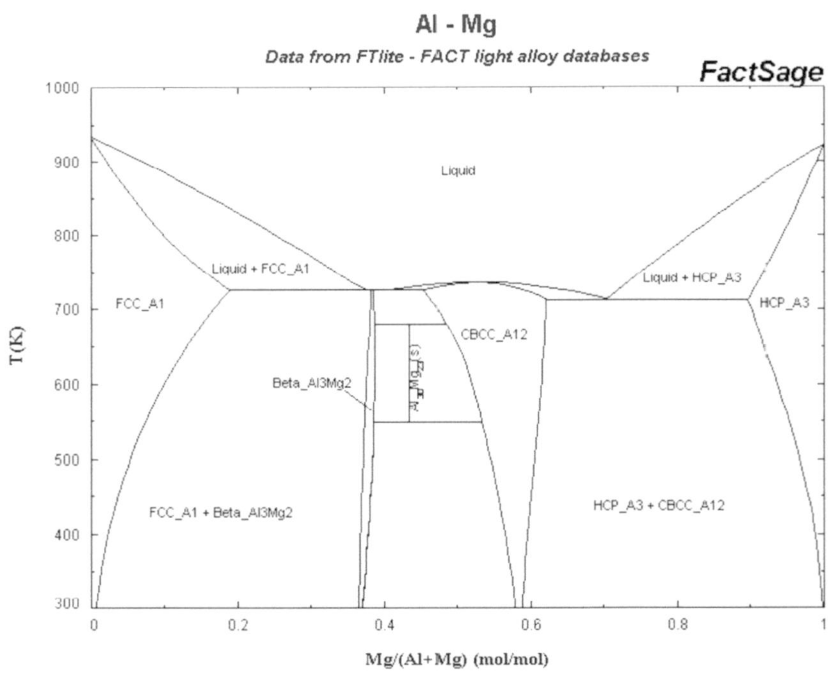

Figure 1. Phase diagram (PD) of binary AlMg alloy [30]. When magnesium in AlMg alloy exceeds its solubility in aluminum liquid, the excessive magnesium will present as intermetallic phases.

http://dx.doi.org/10.5772/intechopen.72763

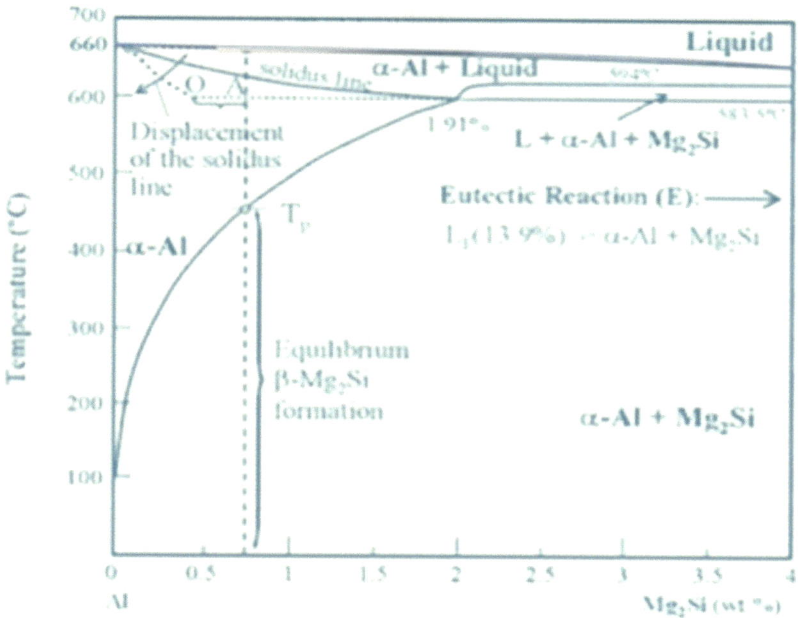

Figure 2. Schematic ternary phase diagram of AlMgSi alloy [31].

precipitates, which possess an ambivalent behavior, by the corrosion process for instance. Alloys on the ternary system AlMgSi with addition of Cu, Fe, Mn and Cr, among others, possess a precipitation sequence where supersaturated solid solution is rich in non-equilibrium vacancies that enable a passage through a series of metastable precipitates towards the equilibrium, described as SSSS → clusters → β″ → β′→ β [5, 6].

3. Experimental details

The 6000 series aluminum alloys containing AlMgSi and AlMgSiCu alloys for such a system are assumed as potential lightweight and high-performance materials to be used in aerospace-craft, aircraft and engine parts in automobiles industry [20]. It has to be accounted for these kinds of systems that the excess of Si in the simplest case induces a major precipitation of MgSi mainly to obtain higher mechanical strength. With this knowledge in mind, and not forgetting that the most important microstructures of solidifying castings underlies on crystallographic arrangement interconnected with physical, chemical, and nature of the atomic alloying. In the face of these facts, the parameters developed for the solidification sequence of AlMgSi alloy have been the cooling rate, chemical nature and concentrations of microalloying elements. Fundamental considerations were taken throughout the molten AlMg liquid solution in order to vary the contact time between the silicon powder added to the aluminum-magnesium molten alloy.

3.1. Melting and solidification process details

Either way, the binary and ternary matrix alloys were processed by stoichiometric ratio of commercially accessible elemental aluminum and magnesium blocks with 99.8 ± 0.1% purity and 10.0 ± 0.02 mm^3 size. The heating and melting elements occur inside of alumina crucible kept in a resistance electrical furnace at 720 ± 5.0°C. This temperature was associated at overheated of the molten AlMg alloy where an interesting and basic oxidation reaction may appear as well as hydrogen dissolution.

With the purpose that compensates certain oxidation losses, two experimental arrangements were considered. The first one was taken in excess 3 wt.% Al and 1.5 wt.% Mg to the required quantity of these constituents. The second consideration was a flux mixture of ionic liquid highly concentrated over the crucible to avoid the contact between metallic melt with the environmental atmosphere and the chemical oxidation. Whiskers of silicon carbide with small diameter of 5.0 ± 0.01 mm were preheated at 300°C and integrated at melt AlMg binary alloy at 1200 rmp via refractory feeder tip directly into the alumina crucible and then was stirring 5–7 minutes for an improved homogenization. In order to have a major role in control of solidification process, the AlMg alloy, firstly, and the ternary AlMgSi melt, secondly, have been poured into a cylinder-shaped iron mold at 100 ± 10°C and superheated above the aluminum liquidus temperature. Stringent conditions, such as high pressures, high vacuum, well-controlled atmospheres or long fabrication times, are not required. Contrary to previous work, the alloys are manufactured under pressure in a squeeze casting [9] to incorporate ceramic particles into aluminum melt generating a homogenate material. From these data, we can draw that our method resolve difficulties like silicon wetting, sinking or floating and viscosity of molten AlMg alloy added with silicon. In addition, our procedure is low cost and relatively readily available [12]. As a consequence, better mechanical properties, microhardness Vickers (MV) particularly, can be achieved by a wider range of silicon addition [13]. This lead on behalf of different solubility limit, associated at each alloying element that is present in the local distribution.

The nominal chemical composition of tested AlMg and AlMgSi alloys obtained by atomic emission spectrometry with a Spectrolab Lax M8-Windows is given in **Table 1**. Here, we would like to highlight certain delicately stoichiometric alloy compositions related to this study with regard to the trace or residual levels of certain chemical elements to achieve a desired set of properties in high-performance AlMg and AlMgSi alloys. The most important recent studies deal with the affinity to certain chemical elements like traces of Fe and P to act

Material alloy	Chemical elements			
	Magnesium	Silicon	Aluminum	Trace elements
AlMg binary	9.58 ± 0.13	0.11 ± 0.01	Bal.	0.79 ± 0.020
AlMgSi ternary	9.71 ± 0.16	2.3 ± 0.017	Bal.	0.81 ± 0.011

Table 1. Chemical composition in wt.% of AlMg and AlMgSi alloys determined by coupled plasma atomic emission spectroscopy.

synergistically to the base alloying elements enhancing the effects of basic alloy constituents. For AlMg alloys, β-$Mg_{17}Al_{12}$ is an essential phase that plays an important role in strengthening crystal boundary. More recent work [28], however, suggests that structural stability mechanism is also explained through the electronic structures. Additional parameter to the trace elements on the alloy phases relates with surface energy in nucleation of the new phase; it is well known that it exerts influence on solubility of components in the binary solutions and contributes to segregation.

3.2. Metallographic analysis

Surface metallographic preparation of the casting AlMg and AlMgSi ingots, cooled at conventional solidification velocity, was obtained for revealing the internal structure arrangement of AlMg and AlMgSi alloys. The physical sectioning at small cylinder specimens was cut with a diamond wheel cutter to gain valuable data about alloys characteristics. Nevertheless, because of the limitation of experimentally detailed solid-liquid interface conditions, the geometry of precision parts close to wall mold produce an accurate model of reality. Due to the surface energy in melts or liquids not varying with direction on the interface, however when the surface of a liquid increased, some atoms come to the surface from the bulk in order to maintain the surface characteristics. For the case of solids, the surface energy differs when the surface is increased, like as the alloys are rolled, as at normal temperatures the atoms mobility is limited [30]. Although the published studies lack data on metallographic and energy surface, this relationship is feasible to resolve the comparative surface energies among crystalline solid phases. From this perspective, the as-cast specimen was carefully encapsulated in epoxy resin, and a basic technique was performed by grinding the samples with SiC from # 400 to # 2000 emery papers and mechanically polished with 1 μm Al_2O_3 power suspension solution in distilled water, followed by etching in Keller's reagent. To prepare the chemical reagent solution, 5 ml of HNO_3 was mixed with 3 ml of HCl, 2 ml of HF and 190 ml of distilled water; the etched was done at room temperature for 30–50 s. Chemical reagent acts on surface sample with remarkable morphology definition due to the energy relative to the surface energy. Taking to account what precedes, the relationship energy fluctuates with temperature. For that case, the temperature coefficient of surface energy is associated to the surface entropy, which varies from entropy in the bulk of the metallic material. The surface entropy is commonly positive. This produces the temperature coefficient of surface energy typically negative. Quantitative and qualitative metallography has significant contributions for the physical surface energy variable.

3.3. Microstructural observations

The morphologies and arrangement of crystalline phases of the as-cast binary AlMg alloy and ternary AlMgSi alloy samples were examined using scanning electron microscopy (SEM) Jeol 5900 with an accelerating voltage of 25 keV. From Leica line-scanning, results of cross section were inspected by electron dispersive spectroscopy (EDS) in order to get the chemical microanalysis and X-ray chemical mapping. A specimen was sectioned at superior, average and inferior part for its chemical and microstructural analyses. These results represent the average of five different analyses in different regions of the surface samples.

In order to know the crystal structural investigation of AlMg and their alloying, a short wave, monochromatic and collimated beam using copper tube of Kα line radiation: $\lambda = 0.15406$ nm of X-rays directed on the carefully polished and cleaned surface of the alloys by a Siemens 5000 X-ray diffractometer. The angle of incidence, θ, is observed, as I is the intensity of the reflected beam graphite monochromator. The crude results are usually presented as a plot of I vs. $(\sin\theta)/\lambda$, where λ is the wavelength of the incident beam recorded from 5 to 120° with step size 0.02 and 0.6 s time per step.

4. Results and discussions

In order to understand the microstructure evolution of AlMg and AlMgSi alloys, we need to analyze their internal structure throughout the growth of complex dendrite. The dendritic structure is an arrangement of primary and secondary arms, where the cavities between the dendrite branches were occupied by eutectic-type phase. As well as simulations, microstructure [32] reveals that the side branches are created by the development of instabilities induced on the tip of the primary branches. Consequently, the morphology of each phase formed has significant effects on the mechanical and material properties of cast alloys. In addition, provide support for a wide range of structural parameters, we may know the phases and transformations to figure out a new emerging uses that provide the AlMg and AlMgSi alloys, which outperform the competition in demanding automotive applications.

Figure 3 shows the chemical composition by EDS of aluminum, magnesium and silicon in the grain boundary regions, which could be related with the diffusion solutes into the matrix alloy. Hence, the chemical distribution of solute magnesium and silicon is vital for development and verification of atomic solidification process. The most important effect of nature and quantity of alloy elements influences the morphology of a solidification reaction that took place under cooling of a solid transformation; moreover, based on chemical composition, impurities and their influence in the melt nucleation state are very complicated with significantly crystallization conditions.

Analyzing the solidification structure using charge materials, melting process and cooling conditions near to practical foundry situation held a certain time that included the relationship in the liquid-solid transition region, distribution of chemical compositions, identification of metastable system or description for forming temperature and composition solidification microstructures. Throughout the atomic content in each region of phases, the alloy provides the chemical influence on corrosion and mechanical properties. In addition, inoculation and even addition of trace elements are associated at thermodynamics reactions on solid and liquid region, which is reason why the EDS analysis is required. The exact significance of nature and content of magnesium and silicon is reveled in the α-Al solid solution with atomic magnesium and silicon dissolved on the gray dark zone. Therefore, this type of alloy obtains solutes like magnesium and silicon distributions associated with the experimental alloy designed and the limit of solubility based on the equilibrium AlMg phase diagram.

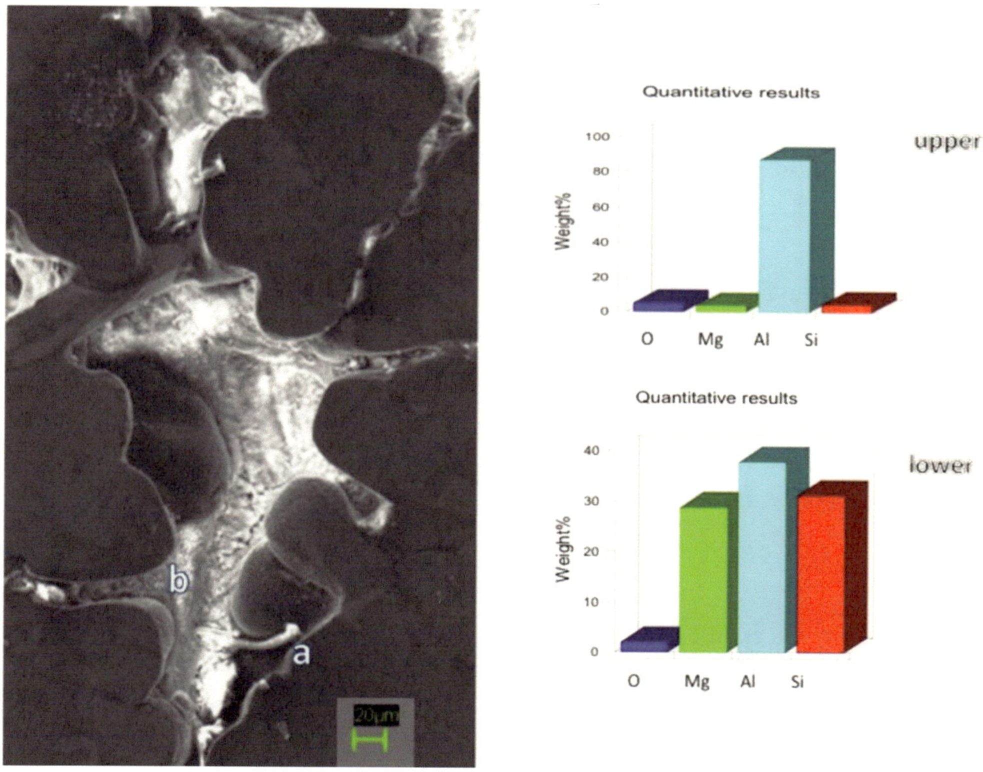

Figure 3. Chemical composition from the AlMg matrix (a) associated with upper graphic bar and interdendritic region (b) that corresponding to lower graphic bar.

The characteristic structural features under the same basic parameters, such as casting temperature and conventional solidification rate, that microstructure features show an equiaxed dendritic morphology observed in a SEM field of back-scattered electrons (BSE) for materials like AlMg and AlMgSi alloys. The microstructure of crystal growth morphology demonstrates the absence of segregation or impurities, and the images of sample surface by secondary electron image (SEI) shown in **Figure 4** did not show internal fissuring, associated with abnormal microsegregation. The electron interactions with the AlMg and AlMgSi atoms produce information about and interdendritic region that can be seen in **Figure 4**. In past years, many studies have reported that microstructure is strongly related with parameters of solidification route, especially by the solidification rate and chemical nature of alloying. In addition, less distance between the walls of mold increase the cooling rate and modified the structure until fine microstructure and less segregation phases. In addition, **Figure 5** shows the crystal structure of AlMg and AlMgSi alloys, which are in agreement with phases reported in equilibrium phase diagram.

In this context, as shown in **Figure 3**, the images by electron microscopy revealed that magnesium and silicon precipitate to a higher extent of interdendritic region. Generally, a uniform distribution of magnesium in the aluminum matrix is observed in the atomic X-ray mapping.

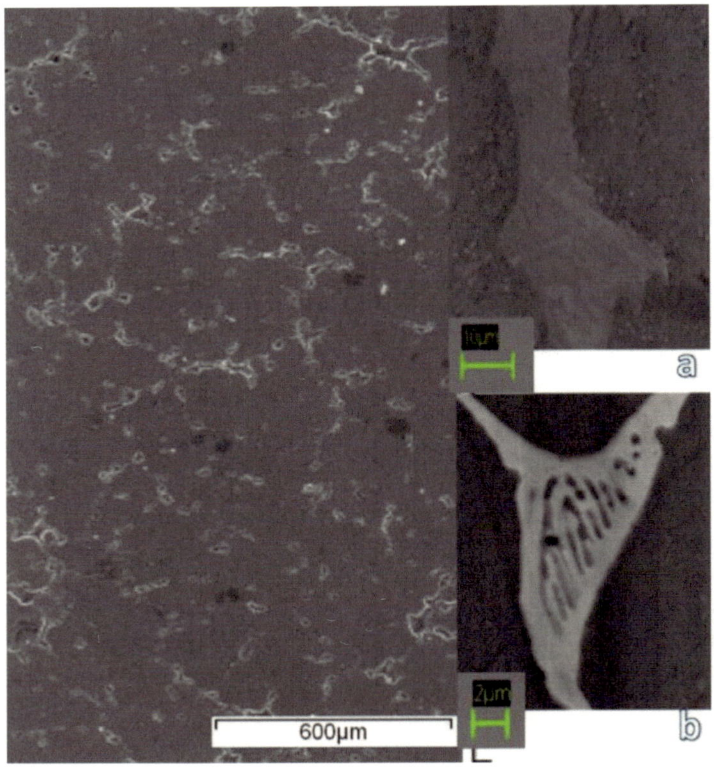

Figure 4. Microstructure of binary AlMg alloy, which was formed by AlMg solid solution (SS) rich in aluminum with dendritic microstructure and revealing the interdendritic morphology.

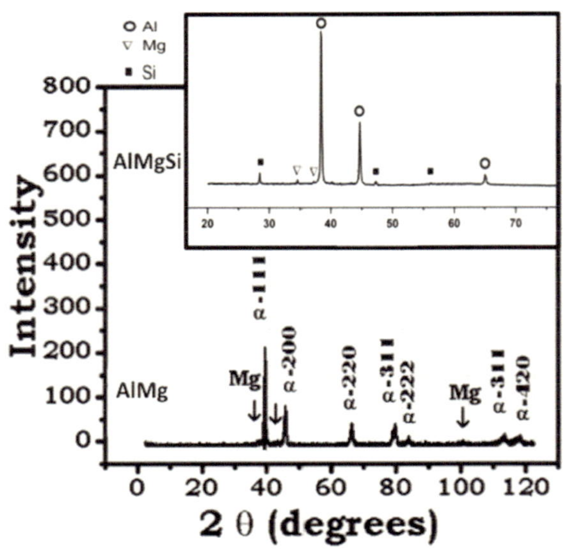

Figure 5. XRD patterns of the AlMg and AlMgSi alloys.

In contrast, silicon is commonly identified in interdendritic regions as shown in **Figure 3**. The structure of crystalline solid, from the phase diagram of AlMgSi ternary alloy [21], shows the content of magnesium and silicon as solid solute alloying in the aluminum-rich corner. It seems, therefore, that solutes comprise a solid solution in aluminum solvent matrix at a specific extent. In this respect, the high concentration of magnesium and silicon could made that the diffusion as interchange of atoms in limit of solid solubility produced precipitation of Mg_2Si phase together with a lower concentration of silicon. However, despite all these metallurgical characterizations, few works concerning the mechanical behavior of these alloys can be found like the research carried out previously [15], where the motion dislocations generated slip in a solid by the motion of dislocations possessing internal strains. However, it should be born in mind that adequate incorporation of grain refinement into the aluminum matrix rise in precipitation due to two physical phenomenal suggestions. The first one deal with rapid heterogeneous nucleation of precipitates, whereas the second one relates an increased rate of dislocation generated near particles added. Characteristically, alloys with precipitates on matrix and refine microstructure age hardening increase the main internal strain and hardness is improved. A further discussion of this fascinating phenomenon of precipitation will be addressed to the corrosion performance, given the fact that the presence of phases along the AlMg matrix may expect a galvanic couple decreasing the corrosion resistance in contact with electrolyte.

5. Conclusions

Summarizing the chapter, it may be stated that numerous phenomena assist solidification process.

A brief mention of microstructure influence has been described in relationship with AlMg and AlMgSi alloys.

The chemical and microstructural details concerning the concentration and distribution of solutes in the unmodified aluminum base alloy were exposed.

Microstructural evidence showed the distribution of phases based on low-cost solidification and basic casting route.

The AlMg and AlMgSi alloys presented Al-rich solid solution matrix with interdendritic regions rich in magnesium and silicon for AlMgSi.

Acknowledgements

This work was done while visit the Engineering School of BUAP. It is a pleasure to acknowledge the fruitful atmosphere during the visit. This research was supported in part by PAPIIT-UNAM, acknowledges support from CONACyT.

Acronyms and Abbreviations

PD	Phase diagram
SEM	Scanning electron microscopy
EDS	Electron dispersive spectroscopy
MV	Microhardness Vickers
SS	Solid solution
XRD	X-ray diffraction
BSE	Back-scattered electrons
SEI	Secondary electron image

Author details

Socorro Valdez Rodríguez[1]*, Lorenzo Martínez[1] and Martín Pech-Canul[2]

*Address all correspondence to: svaldez@fis.unam.mx

1 Institute of Physical Sciences-National Autonomous, University of Mexico ICF-UNAM, University Av. s/n. Cuernavaca, Morelos, México

2 Center for Research and Advanced Studies of the National Polytechnic Institute Cinvestav-Saltillo, Ramos Arizpe, Coahuila, México

References

[1] Xu H, Dang Z. Numerical investigation of coupled mass transport and electrochemical reactions in porous SOFC anode microstructure. International Journal of Heat and Mass Transfer. 2017;**109**:1252-1260. DOI: 10.1016/j.ijheatmasstransfer.2017.02.090

[2] Cuniberti A, Tolley A, Castro Riglos MV, Giovachini R. Influence of natural aging on the precipitation hardening of an AlMgSi alloy. Materials Science and Engineering A. 2010;**527**:5307-5311. DOI: 10.1016/j.msea.2010.05.003

[3] Valdez S, Bautista A, Villanueva M, Pech-Canul MI. Thermochemistry oxidation of trimetallic alloy modified by thermal procedure. International Journal of Innovative Research in Science, Engineering and Technology. 2016;**5**:5235-5242. DOI: 10.15680/IJIRSET.2016.0504131

[4] Valdez S, Campillo B, Pérez R, Martínez L, García A. Synthesis and microstructural characterization of Al-Mg alloy-SiC particle composite. Materials Letters. 2008;**62**:2623-2625. DOI: 10.1016/j.matlet.2008.01.002

[5] Gancarz T, Jourdan J, Gasior W, Henein H. Physicochemical properties of Al, Al-Mg and Al-Mg-Zn alloys. Journal of Molecular Liquids. 2018;**249**:470-476. DOI: 10.1016/j.molliq.2017.11.061

[6] Li H, Cao F, Guo S, Jia Y, Zhang D, Liu Z, Wang P, Scudino S, Sun J. Effects of Mg and Cu on microstructures and properties of spray-deposited Al-Zn-Mg-Cu alloys. Journal of Alloys and Compounds. 2017;**719**:89-96. DOI: 10.1016/j.jallcom.2017.05.101

[7] Zhao Z, Sun Z, Liang W, Wang Y, Bian L. Influence of Al and Si additions on the microstructure and mechanical properties of Mg-4Li alloys. Materials Science & Engineering A. 2017;**702**:206-217. DOI: 10.1016/j.msea.2017.06.077

[8] Zhang J, Ma M, Shen F, Yi D, Wang B. Influence of deformation and annealing on electrical conductivity, mechanical properties and texture of Al-Mg-Si alloy cables. Materials Science & Engineering A. 2018;**710**:27-37. DOI: 10.1016/j.msea.2017.10.065

[9] Kaseem M, Choi K, Ko YG. A highly compact coating responsible for enhancing corrosion properties of Al-Mg-Si alloy. Materials Letters. 2017;**196**:316-319. DOI: 10.1016/j.matlet.2017.03.106

[10] Lucci R, López Padilla R, Cantero S, Bariles R, Oldani C. Refining of AZ91 magnesium alloy obtained in machining chips recycling. Procedia Materials Science. 2015;**8**:886-893. DOI: 10.1016/j.mspro.2015.04.149

[11] Chi Y, Gu G, Yu H, Chen C. Laser surface alloying on aluminum and its alloys: A review. Optics and Lasers in Engineering. 2018;**100**:23-37. DOI: 10.1016/j.optlaseng.2017.07.006

[12] Wan B, Chen W, Lu T, Liu F, Jiang Z, Mao M. Review of solid state recycling of aluminum chips. Resources, Conservation & Recycling. 2017;**125**:37-47. DOI: 10.1016/j.resconrec.2017.06.004

[13] Gangil N, Siddiquee AN, Maheshwari S. Aluminium based in-situ composite fabrication through friction stir processing: A review. Journal of Alloys and Compounds. 2017;**715**:91-104. DOI: 10.1016/j.hallcom.2017.04.309

[14] Anne G, Ramesh MR, Shivananda Nayaka H, Arya SB, Sahu S. Development and characteristics of accumulative roll bonded Mg-Zn/Ce/Al hybrid composite. Journal of Alloys and Compounds. 2017;**724**:146-154. DOI: 10.1016/j.jallcom.2017.07.026

[15] Olbricht J, Yawny A, Young ML, Eggeler G. Mechanical and microstructural observations during compression creep of a short fiber reinforced AlMg metal matrix composite. Materials Science and Engineering A. 2009;**510-511**:407-412. DOI: 10.1016/j.msea.2008.05.057

[16] Prasad LC, Chatterjee SK, Singh VN. Intermetallic associations in AlMg liquid alloys. Physica B: Condensed Matter. 1996;**217**:283-291

[17] Ajmera D, Panda E. Thermodynamics of ultra-thin oxide overgrowths on Al-Mg alloys: Role of interface energy. Corrosion Science. 2016;**102**:425-436. DOI: 10.1016/j.corsci.2015.10.035

[18] Tanski T, Snopinski P, Pakiela W, Borek W, Prusik K, Rusz S. Structure and properties of AlMg alloy after combination of ECAP and post-ECAP ageing. Archives of Civil and Mechanical Engineering. 2016;**16**:325-334. DOI: 10.1016/j.a cme.2015.12.004

[19] Umaru OB, Abdulwahab M, Tokan A, Bello AM, Umar HA. Effect of double thermal ageing treatment on the mechanical properties of Al-Cu-Mg/3% rice husk ash composite. Results in Physics. 2016;**6**:342-345. DOI: 10.1016/j.rinp.2016.06.009

[20] Miyake Y, Sato Y, Teranishi R, Kaneko K. Effect of heat treatments on the microstructure and formability of Al-Mg-Mn-Sc-Zr alloy. Micron. 2017;**101**:151-155. DOI: 10.1016/j.micron.2017.07.003

[21] Vlach M, Cizek J, Smola B, Melikhova O, Vlcek M, Kodetová V, Kudrnová H, Hruska P. Heat treatment and age hardening of Al-Si-Mg-Mn commercial alloy with addition of Sc and Zr. Materials Characterization. 2017;**129**:1-8. DOI: 10.1016/j.matchar.2017.04.017

[22] Choi SW, Cho HS, Kang CS, Kumai S. Precipitation dependence of thermal properties for Al-Si-Mg-Cu-(Ti) alloy with various heat treatment. Journal of Alloys and Compounds. 2015;**647**:1091-1097. DOI: 10.1016/j.jallcom.2015.05.201

[23] Patakham U, Kajornchaiyakul J, Limmaneevichitr C. Grain refinement mechanism in an Al-Si-Mg alloy with scandium. Journal of Alloys and Compounds. 2012;**542**:177-186. DOI: 10.1016/j.jallcom.2012.07.018

[24] Takai R, Kimura S, Kashiuchi R, Kotaki H, Yoshida M. Grain refinement effects on the strain rate sensitivity and grain boundary sliding in partially solidified Al-5wt%Mg alloy. Materials Science & Engineering A. 2016;**667**:417-425. DOI: 10.1016/j.msea.2016.05.023

[25] Gariboldi E, Bassani P, Albu M, Hofer F. Presence of silver in the strengthening particles of an Al-Cu-Mg-Si-Zr-Ti-Ag alloy during severe overaging and creep. Acta Materialia. 2017;**125**:50-57. DOI: 10.1016/j.actamat.2016.11.056

[26] Wang Y, Luo G, Zhang J, Shen Q, Zhang L. Effect of silver interlayer on microstructure and mechanical properties of diffusion-bonded Mg-Al joints. Journal of Alloys and Compounds. 2012;**541**:458-461. DOI: 10.1016/j.jallcom.2012.06.120

[27] Nanjo F, Kobayashi J, Itoh G. Spinning workability of Al-Mg-Si alloy extruded tube using the forming die. Procedia Engineering. 2017;**207**:1761-1766. DOI: 10.1016/j.proeng.2017.10.935

[28] Remoe MS, Marthinsen K, Westermann I, Pedersen K, Royset J, Marioara C. The effect of alloying elements on the ductility of Al-Mg-Si alloys. Materials Science & Engineering A. 2017;**693**:60-72. DOI: 10.1016/j.msea.2017.03.078

[29] Razavykia A, Farahany S, Yusof NM. Evaluation of cutting force and surface roughness in the dry turning of Al-Mg_2Si in-situ metal matrix composite inoculated with bismuth using DOE approach. Measurement. 2015;**76**:170-182. DOI: 10.1016/j.measurement.2015.08.032

[30] Valdez S, Chigo-Anota E, Pech Canul MI, Juárez-Islas JA. Degradation analysis of aluminum cellular alloy in saline ionic electrolyte. International Journal of Innovative

Research in Science, Engineering and Technology. 2015;**4**:10337-10345. DOI: 10.15680/IJIRSET.2015.0411004

[31] Voort GV, Suárez-Peña B, Asensio-Lozano J. Microstructure investigations of streak formation in 6063 aluminum extrusions by optical metallographic techniques. Microscopy and Microanalysis. 2013;**19**:276-284. DOI: 10.1017/S143192761300010X

[32] Liu CL, Azizi-Alizamini H, Parson NC, Poole WJ, Du Q. Transactions of Nonferrous Metals Society of China. 2017;**27**:747-753. DOI: 10.1016/S1003-6326(17)60085-2

Chapter 5

Solidification Versus Adsorption for Immobilization of Pollutants in Geopolymeric Materials: A Review

Bassam I. El-Eswed

Additional information is available at the end of the chapter

http://dx.doi.org/10.5772/intechopen.72299

Abstract

Geopolymer (GP) is a class of three-dimensional aluminosilicate binder, which is superior to Portland cement materials in acid, heat and fire resistance. GP is produced by reacting an aluminosilicate source (metakaolin, fly ash or waste) with an alkali metal hydroxide or silicate. The aim of the present work is to review the latest developments in three lines of research that deal with application of GP in treatment of pollutants. The first **"intra-solidification"** that involves mixing real waste (containing heavy metal pollutants) with the GP precursors to obtain a high mechanical strength material. The second type of solidification is **"inter-solidification"** that involves incorporation of heavy metals solutions (as simulation of polluted water) during geopolymerization reaction. The third line of research **"adsorption"** involves agitating GP with heavy metals solutions and studying the ability of GP to remove heavy metals from water. These techniques will be investigated regarding efficiency and mechanism of immobilization, cost and environmental impact. GPs are strong low-cost adsorbents for heavy metals. In intra-solidification, despite the high mechanical strength of the produced GP-containing waste, geopolymerization reduces effectively the leaching of heavy metals. The reverse was observed in the case of inter-solidification which presents a greater challenge than intra-solidification.

Keywords: solidification, adsorption, geopolymer, heavy metals, fly ash

1. Introduction

Geopolymers (GPs), first named by Davidovits [1], are synthetic aluminosilicates prepared by reacting a reactive solid aluminosilicate (metakaolin, fly ash or blast furnace slag) with an aqueous alkaline activating solution that contains alkali (Na^+ or K^+) hydroxides and silicates [2, 3]. GPs are composed of tetrahedral silicate and aluminate units linked in a three-dimensional structure by covalent bonds. The negative charges associated with Al(III) tetrahedra is usually balanced by alkali cations [4]. GPs possess many favorable properties such as high compressive strength, low shrinkage, acid resistance, fire resistance, low thermal conductivity, rapid hardening and durability [2, 4–6]. The main applications of GPs are low cost construction materials and protective coatings [6]. Furthermore, GPs have been used for solidification of toxic wastes especially those containing heavy metals. This is the first aspect that will be reviewed in the present work. Solidification is a pre-landfill waste treatment in which the waste is encapsulated in cementing material or any kind of organic or inorganic polymer to immobilize toxic contaminates and reducing their leachability [7]. The final aim of solidification is reduction of the overall environmental impact of waste disposal [8].

GPs are X-ray amorphous, however, the formation of some crystalline zeolitic phases was detected particularly. The transformation of typically X-ray amorphous GP synthesized in highly concentrated basic solution into zeolite structure provides a very powerful evidence for the similarity of the zeolitization and geopolymerization reaction mechanisms [9]. Zeolites are strong cation exchangers and were used extensively to remove heavy metal ions from polluted water. Thus, this similarity suggested using GP in water treatment for the removal of heavy metal ions. Adsorption of pollutants, especially heavy metals, on GP is the second aspect that will be discussed in the present work.

Any material that is rich in Si and Al in amorphous forms can be used as a source material for synthesis of GP. Natural resources like kaolin and metakaolin (MK, synthesized by dehydroxylation of phase pure kaolin at 750°C) can be used for GP preparation, as well as a considerable range of industrial waste, such as fly ash (FA) from coal, blast furnace slag (BFS) from metallurgical industries, and municipal and medical waste incineration ashes. Since GP production is mainly based on waste materials, their use could aid waste minimization and reduces the pressure on current and future limited resources used in ordinary Portland cement (OPC) and contribute significantly to the lowering of the CO_2 emission of the construction industry [2, 6]. Calcination of kaolin to produce MK is carried out at 750°C, which is lower than the temperature needed in OPC production (1400–1450°C), [4]. Coal fly ash (FA) is an industrial waste that is produced from coal electricity plants. The bulk of the FA is made up of silicon, aluminum and iron oxides, as well as small (type F) or significant amounts of CaO (type C) [9]. GPs are generally considered to be much more sustainable than OPC, in terms of reduced production energy requirement, lower CO_2 emissions and the inexhausting sources of wastes [4]. FA appears to be one of the most promising precursors for the production of GP due to its high workability, low water demand and high mechanical strength [2, 9].

Despite the typical attributes mentioned above, the properties of GP are highly dependent on Si/Al ratio, hydroxide concentration, alkali cation used and curing conditions [6]. For example, the compressive strength of MK-based GP was improved by using sodium or potassium silicate solution instead of hydroxide. This improvement is attributed to the presence of increased number of Si—O—Si bond, which possess higher bond energy compared to Si—O—Al bond [5]. It was observed that the microstructure changed from containing large to small pores as the Si/Al ratio was increased from 1.15 to 1.90 [9, 10]. The baseline stoichiometries for GP materials participating in the reaction are typically molar Na/Al = 1 and Si/Al = 2. The amount of water (H_2O/Na molar ratio ~7) is usually controlled to obtain a thick paste [11].

The mechanism of geopolymerization is out of the scope of the present article, but some points are discussed here. Dissolution of amorphous aluminosilicates (from MK, FA or BFS) is rapid at the high pH provided by the alkaline activator, and this quickly creates a supersaturated solution containing a complex mixture of silicate, aluminate and aluminosilicate species. Then the oligomers in the aqueous phase form large networks by condensation resulting in gel formation. After gelation, the system continues to rearrange and reorganize, as the connectivity of the gel network increases, resulting in the three-dimensional aluminosilicate network of GP. The Si(IV) and Al(III) in the framework of GP are tetrahedrally coordinated and linked by oxygen bridges with small amount of terminal hydroxyl groups at the surface of the gel [9]. Although these terminal hydroxyl groups were thought to be insignificant in determining the mechanical properties of GP, they were thought to be important in immobilization of heavy metals since they can be coordinated to the heavy metal cations through chemical bonds [12].

The aim of the present work is to review the latest developments in the techniques of immobilization of heavy metals and some organic pollutants in geopolymeric materials. Two techniques will be investigated and compared; the first is solidification and the second is adsorption. These will be reviewed regarding starting materials used for preparation of GP, efficiency, mechanism, cost and environmental impact.

2. Methodology and calculations

2.1. Solidification

The first part of the present study is devoted to reviewing published works on solidification of wastes containing heavy metals in GP matrix. This type of solidification, which will be called **intra-solidification** (**Figure 1**), involves mixing the waste with aluminosilicate-rich precursor such as MK, FA and BFS. Then an activator composed from sodium or potassium hydroxide and silicate is added and the produced GP is cured at temperature usually ranges from 40 to 80°C. The prefix "intra" is used because it means "within" and thus intra-solidification means immobilization of the heavy metals within waste by reaction with GP precursor. The second type of solidification, which will be called **inter-solidification** (**Figure 1**), involves mixing an aluminosilicate precursor with pure heavy metal salts or solutions added during preparation of GP. The prefix "inter" is used because it means "between" and thus inter-solidification means immobilization reaction between heavy metal added and the GP precursors.

Two criteria are usually used to test the efficiency of solidification, the first is mechanical strength of the GP and the second is the leaching of heavy metals from the GP. Toxicity characteristic leaching procedure (TCLP) formulated by the US Environmental Protection Agency (EPA) involves stirring granular solid GP material (containing heavy metal) < 9 mm with acetic acid solution at pH 3–5 (liquid:solid ratio 20 mL:1.0 g) followed by filtering and determining the amount of heavy metals [8, 13]. The European Norm EN-12457 test, which is less aggressive than TCLP test, involves immersing 4.0 g of GP in 40 mL deionized water [8, 13, 14]. Furthermore, other leaching solutions of different pH and ionic strength were applied by other authors: H_2SO_4 solutions (pH 1), 5% (w/w) $MgSO_4$, 5% Na_2CO_3 solutions [15], 0.1 M HCl, 0.1 M NaCl, 1.0 M NaCl and 0.1 M NaOH [12]. The time of leaching test is highly variable and

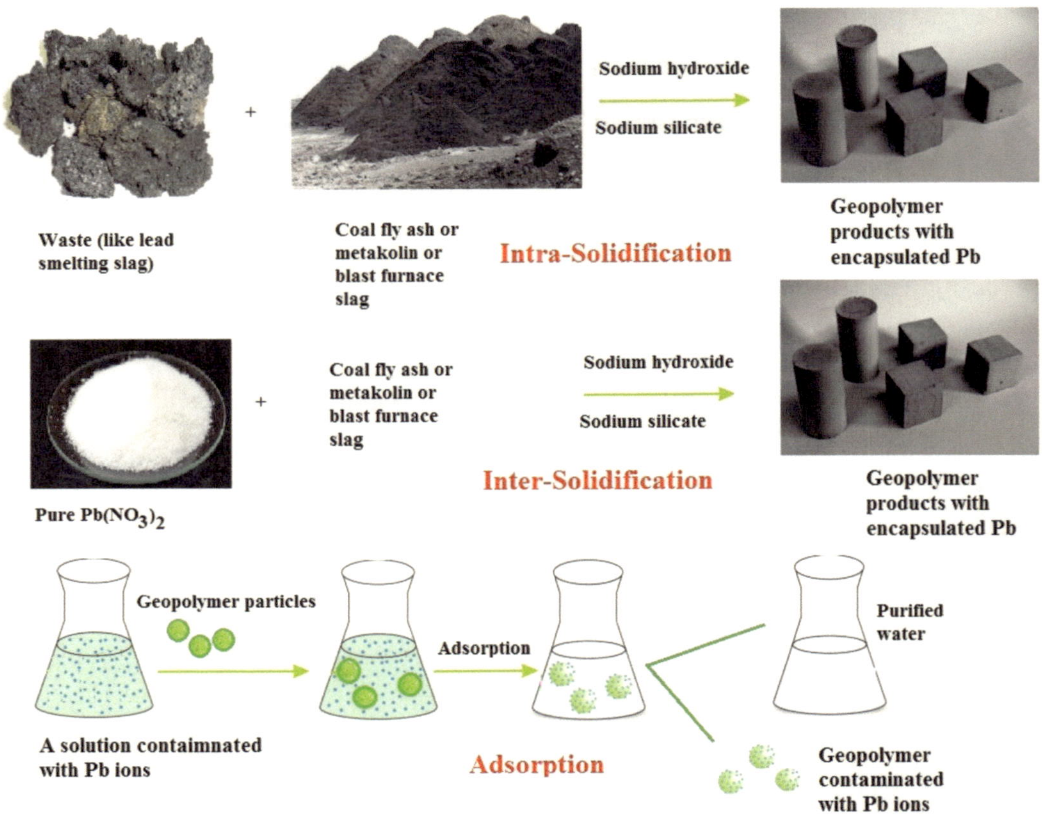

Figure 1. Intra-solidification, inter-solidification and adsorption processes for immobilization of heavy metals.

ranges from 1 to 90 days. The efficiency of solidification can be deduced from the concentration of metal in the leaching solution when compared with the allowed limit formulated by environmental organizations. If the concentration of heavy metal in the leaching solution is below the allowed limit, then it can be concluded that the solidification process is effective.

Another way for evaluating the efficiency of solidification depends on calculating the percentage of leaching of heavy metal from GP containing the heavy metal using Eq. (1) [12]:

$$\%\,\text{leaching of heavy metal} = \frac{C_{GPM}}{C_t} \times 100 \tag{1}$$

where C_{GPM} is the concentration of heavy metal leached out of the GP-containing heavy metal (mg/L). C_t is the calculated concentration (mg/L) of heavy metal (Eqs. (2) or (3)) expected to leach out of specific mass (m, in mg) of GP into specific volume (V in liters) of leaching solution assuming 100% leaching:

$$\text{Intra} - \text{solidification: } C_t = \frac{mass\ of\ heavy\ metal\ (g)}{mass\ of\ waste\ (kg)} \times \frac{mass\ of\ waste\ (g)}{mass\ of\ GP\ (g)} \times \frac{m}{V} \tag{2}$$

$$\text{Inter} - \text{solidification: } C_t = \frac{mass\ of\ metal\ salt\ used\ in\ mix\ design\ (g)}{mass\ of\ mix\ desigh\ of\ GP\ (g)} \times \frac{m}{V} \times \frac{atomic\ mass\ of\ heavy\ metal}{molar\ mass\ of\ heavy\ metal\ salt} \quad (3)$$

It is worth to mention that the amount of heavy metal reported in the waste material used in Eq. (2) (in g/kg, **Table 1**) is problematic because semi-quantitative X-ray fluorescence (XRF) gives the amount of metal that exist in different forms. Some of these forms may be inert. For example, as shown in **Table 1**, the environmentally significant amount of Pb in lead smelting slag determined by digesting in 2 M HNO_3 (14.7 g/kg) was much lower than the amount indicated by X-ray fluorescence (319.3 g/kg) [14]. Furthermore, many authors reported the % metal in GP total mix design incorrectly and did not differentiate between % metal and % metal salt. For example, 0.50% $Cu(NO_3)_2{\cdot}3H_2O$ is equivalent to 0.13.% Cu.

2.2. Adsorption

GP was used as an adsorbent for the removal of heavy metals from polluted water (**Figure 1**). Adsorption involves agitating specific amount of GP (m, in g) with specific volume (V in L) of standard aqueous solution (C_i) of heavy metal (simulating polluted waste water) for different contact times. The concentration of heavy metal in solution decreases until it reaches steady value (C_e) at a time called equilibration time. The mount of heavy metal adsorbed at equilibrium (Q_e, mg/g) is determined from Eq. (4):

$$Q_e = \frac{C_i - C_e}{m} \times V \quad (4)$$

Langmuir model (Eq. (5)) is the most widely used in the studies reviewed in the present article. This model is used to fit the equilibrium adsorption data that results from determining C_e and Q_e at different C_i values at constant temperature (adsorption isotherm).

$$Q_e = \frac{Q_m \times K_L \times C_e}{1 + (K_L \times C_e)} \quad (5)$$

where Q_m is the maximum adsorption capacity (mg/g) and K_L is the Langmuir affinity constant related to the energy of binding sites on the surface of GP (L/mg) [16].

For analysis of kinetics adsorption data, pseudo-second order model (Eq. (6)) was the most frequently used in the works reviewed in the present article:

$$\frac{dQ_t}{dt} = k_2(Q_e - Q_t)^2 \quad (6)$$

where k_2 is the second-order reaction constant (g mg^{-1} min^{-1}), Q_e and Q_t are the amount of metal ions adsorbed per unit mass of GP at equilibrium and time t, respectively. The model is usually used to determine the rate constant k_2 [17].

Waste [Ref.]	Zn	Cd	Cr	Pb	Cu	Ni	As	Mo	Co	Ba	Sn	Hg	Sb	Se	V	Sr
Carbon steel electric arc dust [8, 49]	342.9	0.29	2.108	2.29												
	260	0.8	9.5	13.0		2.0	2.9									
Lead smelting slag [14]	1.690*	<0.04	0.470	319.3	0.760	0.410	2.580	0.180	<0.037	1.970	4.25		11.0	0.420	0.070	<0.04
	0.770**	0.013	0.0035	14.7	0.437	0.034	0.362	0.005	0.059	0.039			0.171	0.008	0.010	0.012
Waste solution after plating of printed circuit board [27]	0.0061		0.0026	12.619	0.3475	0.0002					2.371					
Municipal solid waste incineration fly ash [28]		0.105	0.196	0.502	0.473	0.038										
Municipal solid waste incineration fly ash [29]	3.692	0.103	1.785	0.826	2.817	0.078										
Municipal solid waste incineration fly ash [13]	15.9	0.006	0.790	0.398	0.156	0.090	0.060	0.046	0.094	1.270	1.79	0.060	5.11	0.080	0.040	
Municipal solid waste incineration fly ash [30]	3.269	0.0367	0.157	1.515	0.563	0.071							0.0358			

*Semiqualitative XRF.
**Environmentally significant available after 2 M HNO_3 digestion.

Table 1. Heavy metal content of different wastes (g/kg).

2.3. Cost calculations

The cost of GP used in solidification and adsorption of wastes was calculated depending on the cost values of the constituents of GP given by McLellan et al. [18]. The approximate cost of MK, FA, sodium hydroxide, sodium silicate solution, kaolin, zeolite and BFS used in calculation was 200, 50, 400, 200, 100, 100 and 50 US $/t, respectively. Water and industrial waste solidified were assumed to have negligible cost if solidification is to be applied in the site of production of waste.

3. Immobilization of pollutants in geopolymeric materials

3.1. Intra-solidification of waste in geopolymer

Heavy metals exist in many types of wastes: coal fly ash, carbon steel electric arc furnace dust, lead smelting slag, chromite ore-processing residue, aluminum production waste, municipal solid waste incineration fly ash and medical waste incineration fly ash. The heavy metal content in some of these wastes was given in **Table 1**. Because of their toxic heavy metals content, regulations prohibit disposal of these wastes in landfill [19]. For solidification of these wastes in GP matrix, these wastes are usually mixed with MK, FA or BFS and then with alkali activator to get solid GP after curing.

3.1.1. Coal fly ash waste

Bankowski et al. investigated solidification of brown coal type C fly ash {30.6, 2.4, 1.2} in MK {0.1, 49.3, 35.0}-based GP using sodium hydroxide and sodium silicate as activators. The TCLP test (**Table 2**, I-1) indicated that geopolymerization reduces the leaching of some metals such as As, Ba, Se and Sr (0.002, 0.07, 0.130 and 5.500 mg/L) relative to the unreacted fly ash (0.012, 0.270, 0.740 and 31.200 mg/L). Other metals which exist in trace amounts (like Zn, Mn, Ni and Mo) in fly ash did not respond to solidification due to error in measurements of low concentrations of these metals. It is worth to mention that best immobilization efficiency was obtained when ≤40% (w/w) of the fly ash was incorporated in the mix design of GP [19]. Similarly, Arioz et al. studied solidification of FA (F-type, low calcium) in GP without addition of any other source of aluminosilicate and using sodium hydroxide and silicate activators. TCLP leaching results indicated that heavy metals like As and Hg were immobilized effectively in GP while Zn, Pb and Cd were not (**Table 2**, I-2). The compressive strength of the produced GP ranged from 20 to 50 MPa [20]. The results in **Table 2** (I-1 and I-2) indicated that leaching of both unreacted FA and solidified FA-GP gives low concentrations of heavy metals.

3.1.2. Metal industry waste

Solidification of 20% (w/w) carbon steel arc furnace dust {8.60, 6.21, 2.0} in OPC and type-F FA {3.94, 63.91, 21.51}-based GP was studied using variable activators (sodium and potassium hydroxides and silicates). The compressive strength of the produced GP ranged from 10 to 40 MPa compared to 2.5 MPa in the case of OPC. TCLP leaching test (acetic acid buffer at pH 3) indicated that the leached amounts of Zn, Pb, Cr and Cd from OPC system were less than 1%

Waste [Ref.]	Host matrix	As	Ba	Cr	Cu	Mn	Mo	Ni	Se	Sr	V	Zn	Cd	Pb	Co	Sn	Hg	Sb
Brown coal fly ash (precipitator) [19] I-1	Unreacted*	0.012	0.270	0.004	0.018	0.010	0.011	0.026	0.740	31.20		0.038						
	Solidified in MK-GP*	0.001–0.002	0.07–0.12			0.140–0.28	0.012–0.017	0.09–0.16	0.06–0.13	1.1–11.5	0.010–0.130	0.030–0.070						
Fly ash (F-type) [20] I-2	Unreacted*	0.40		0.058								<0.002	<0.002	<0.01			0.394	
	Solidified in FA-GP*	<0.0004		0.0082–0.023								0.0015–0.035	0.002–0.009	0.008–0.011			<0.003	
Carbon steel arc furnace dust [8] I-3	Solidified in FA-GP*			0.005–0.09								508–1296	1.2–4.52	5.5–31.7				
	Solidified in OPC*			2.7								0.1	0.03	0.6				
Carbon steel arc furnace dust [21] I-4	FA-GP*			<0.05								500–1300	1–4.5	7–32				
	OPC*			2.7								0.1	0.03	0.6				
Lead smelting slag [14] I-5	Unreacted*	0.10–0.13	1.49–1.77	0.05–0.09								0.21–0.28	16.6–37.0					
	Solidified in FA/BFS-GP*	0.01–0.06	1.75–2.84	<0.0006								0.01–0.03	4.8–23.5					
Chromite processing residue [26] I-6	Unreacted*			67.89														
	Solidified in BFS-GP*			0.027–1.55														
Waste solution after plating of printed circuit boards [27] I-7	Solidified in MK/BA-GP*			0.046–0.083	0.08–1.0									0.045–0.160		0.05–0.20		
Municipal solid waste incineration fly ash [28] I-8	Unreacted**			2.31	2.63			0.00					3.22	3.94				
	Solidified in MK-GP**			0.64 (100)	0.07 (1.8)			0.00					0.00	0.03 (0.7)				

Waste [Ref.]	Host matrix	As	Ba	Cr	Cu	Mn	Mo	Ni	Se	Sr	V	Zn	Cd	Pb	Co	Sn	Hg	Sb
Municipal solid waste incineration fly ash [29] I-9	Solidified in FA-GP***			9.8–14.3 (11–16)	12.1–18.3 (9–13)							1.2–2.2 (0.7–1.2)						
Municipal solid waste incineration fly ash [13] I-10	Solidified in FA/MK-GP*	≤0.03	0.12–0.61 (0.2–1)	0.51–0.73 (1–2)	≤0.01		0.2–0.4 (10–17)	0.04–0.18	≤0.01		0.13–0.26 (7–13)	0.70–9.8 (0.9–1.2)	≤0.01	≤0.03	0.055–0.13	≤0.02	≤0.03	0.02–0.43
	Solidified in OPC*	≤0.03	0.24–0.58 (0.2–0.4)	0.21–0.52 (0.5–1.3)	≤0.01		0.18–0.43 (8–19)	≤0.01	≤0.01		0.02–0.1 (1–5)	≤0.01	≤0.01	≤0.03	≤0.01	≤0.02	≤0.03	0.18–1.4
Municipal solid waste incineration fly ash [30] I-11	Solidified in FA-GP*			0.16 (2.04)	0.04 (0.14)							0.02 (0.01)		<0.05 < (0.06)				
Medical waste incineration fly ash [7] I-12	BA unreacted*		2.5	0.34								0.15		0.01				
	FA unreacted*		1.8	0.19								11.8		6.0				
	Solidified in MK-GP*		0.25–0.99	0.025–0.73								0.012–2.15						
USEPA TCLP allowed limit (mg/L)		5	100	5					1			300	1	5			0.2	

*Leaching tests were according to USEPA TCLP procedure.
**Leaching test was according to European EN 12457 procedure.
***Leaching test was done using HNO_3 solution, pH 4, liquid to solid ratio 20 mL/g.
MK: metakaolin, FA: fly ash, BA: bottom ash, BFS: blast furnace slag.

Table 2. Leaching of heavy metals (mg/L) from geopolymer matrix containing solidified waste—intra solidification (values in brackets give calculated % leaching).

of the amount leached from GP. Furthermore, the amounts of metals leached from GP were higher than USEPA limit (**Table 2**, I-3, I-4). This was explained by the high alkalinity of OPC leachate (pH 11.2–11.8) compared to neutral pH of GP leachate (pH 5.2–6.7). However, less aggressive EN leaching test (deionized water) showed much lower leaching concentration for GP samples and much higher leaching concentrations for OPC [8, 21]. This reflects the effect of leaching solution on efficiency of immobilization.

Ogundiran et al. studied solidification of lead smelting slag in (2, 1 w,w) FA (6.03, 48.9, 27.8}-and BFS {42.0, 37.2, 11.8}-based GP using potassium silicate as activator. The compressive strength was remarkably high 80–100 MPa. TCLP leaching test indicated that Pb, Zn, Cr and Ba leaching concentrations were below the recommended limit (**Table 2**, I-5). Many mechanisms were suggested for solidification of Pb in GP. The first, Pb is a network forming element replacing Si, the second, Pb(II) ions balance the negatively charge Al tetrahedra and the third, insoluble $Pb(OH)_2$ be encapsulated in the GP structure [14].

Perna and Hazlicek solidified semi-liquid red waste {24.76, 13.45, 13.88} and gray slag {1.17, 0.01, 54.55} from aluminum production in GP matrix. The geopolymerization of the low silicate waste was carried out with addition of silicon-rich fired clay dust {0.14, 50.28, 4.99} and potassium silicate activator to obtain SiO_2/Al_2O_3 and K_2O/Al_2O_3 molar ratio of 2.96 and 0.73, respectively. The mechanical strength of the produced GP was 30–40 MPa. Foamed GP products with thermal conductivity 0.169 Wm^{-1} K^{-1} were prepared. The high porosity of the foamed GP could be used in insulation and passive house construction and it has fire and elevated temperature resistance advantage over polystyrene [22].

Salihoglu et al. investigated solidification of 25% (w/w) antimony waste slag in FA {4.65, 51.84, 24.68}-based GP with OPC, clay, BFS and gypsum additives and using sodium hydroxide and silicate activators. The EN leaching concentration ranged from 0.83 to 3.29 mg/L in the case of Arsenic (As) and from 3.91 to 8.10 mg/L in the case of antimony (Sb). These concentrations were higher than the regulatory limit 2.5 and 0.5 mg/L in the case of As and Sb, respectively [23].

Rao and Liu reviewed literature (dated to 2015) on the potential of using geopolymerization reactions in solidification of mine and oil sands tailings which contain aluminosilicate materials. The review showed conflicting conclusions in the geopolymerization processes regarding for instance the suitable Na/Al and Si/Al ratios, the function of metal cations such as sodium and potassium, and the effect of calcination (for removal of organic content) [24].

Chromite ore-processing residue (COPR) was a target for solidification in several studies. Sun et al. solidified 12% (w/w) COPR {34.27, 6.46, 8.15} in MK {0.05, 46.58, 37.02}-based GP in the presence of Na_2S (reducing agent) and using sodium hydroxide and silicate activators. The mechanical strength of the produced GP was 40–60 MPa. The TCLP leaching test indicated that the concentration of leached Cr from GP was below the allowed limit (5 mg/L) when the S^{2-}/Cr molar ratio > 6 compared to 45 mg/L when Na_2S was not used. The effect of geopolymerization on the efficiency of solidification was remarkable because the leaching of total Cr from of unreacted COPR was 279.84 mg/L. The suggested mechanism of solidification was that S^{2-} anion reduces anionic Cr(VI) in COPR to cationic Cr(III) which balances the negatively charged Al tetrahedra in the GP [25]. Huang et al. studied solidification of COPR {0.69, 6.19,

11.4} in BFS {40.43, 10.77, 30.18}-based GP using sodium hydroxide and silicate activators. The mechanical strength was found to decrease from 50.4 to 13.4 MPa and the TLCP leaching concentration increases only from 0.027 to 1.553 mg/L (**Table 2**, I-6) when the % (w/w) COPR in GP increased from 10 to 60%. This indicated that 60% (w/w) COPR could be effectively solidified with leaching concentration of Cr below the recommended limit (5 mg/L) [26].

Antunes Boca Santa et al. solidified the waste solution of plating printed circuit boards in MK {0.06, 47.4, 37.9} and bottom ash {1.71, 57.3, 23.7} GP, and using sodium hydroxide and silicate activators. The mechanical strength of the produced GP ranged from 5 to 25 MPa. The concentrations of metals in the TCLP leachate were below the recommended limit for Pb, Cu, Cr, Sn, Fe, Ni and Zn (**Table 2**, I-7). The results indicated that increasing the concentration of NaOH activator from 8 to 12 M causes increase in the concentration of heavy metals in the leachate and thus reduces the efficiency of solidification [27].

3.1.3. Municipal solid waste incineration fly ash

Lancellotti et al. studied solidification of 20% (w/w) of municipal solid waste incineration fly ash {10.46, 5.63, 2.10} in MK-based GP using sodium hydroxide and sodium silicate activators (Si/Al and Na/Al molar ratios of 2 and 1, respectively). EN leaching test revealed that the release of heavy metals from GP is low compared to unreacted waste, consequently, indicating efficient solidification (**Table 2**, I-8). The % leaching for Cu, Pb and Cd was 1.8, 0.7 and 0%, respectively. The effective immobilization of these heavy metals was ascribed to formation of metal hydroxides in the high alkalinity medium of GP preparation. On the other hand, the failure in Cr immobilization was ascribed to the effect of high Cl content of fly ash (26.40%) which promotes dissolution of Cr [28].

Zheng et al. investigated the effect alkaline dose of sodium hydroxide and silicate (Na/fly ash and Si/Al molar ratios) on geopolymerization of municipal solid waste incineration fly ash {28.8, 15.4, 7.2} without addition of extra aluminosilicate source. Using HNO_3 leaching solution (pH 4 with liquid/solid ratio 20), the % leaching of Cr, Zn, Cu was low: 11–16, 0.7–1.2 and 9–13, respectively (**Table 2**, I-9). However, the concentrations of metals in the leaching solution were higher than the TCLP allowed limits. The high Si/Al molar ratio (2.5) was found to reduce leaching of Cr and Cu because this ratio decreases depolymerization during leaching since Si-O-Si bond has higher strength than Si-O-Al bond. An intermediate alkaline dosage (2.4 mol Na/kg FA) favored immobilization of Cr, Cu and Zn [29].

Luna Galiano et al. studied solidification of municipal solid waste incineration fly ash {44.7, 26.8, 12.0} in GP prepared using FA (type F) mixed with kaolin or MK, BFS and different sodium and potassium hydroxides and silicates activators. The leachate pH (TCLP test) was inferred to be the most important variable on immobilization of metals. The results indicated that Zn, Co, Ni, Cu and Sn were the best immobilized, while elements forming oxyanions like Sb, Mo, V and Cr were the least immobilized in GP (**Table 2**, I-10) [13].

Guo et al. solidified 10% (w/w) municipal solid waste incineration fly ash {30.7, 17.7, 4.59} in type C fly ash {13.6, 40.2, 17.0}-based GP using sodium hydroxide and silicate as activators to obtain Na_2O/SiO_2 molar ratio of 0.7. The compressive strength of the produced GP was 50 MPa and the leaching concentrations of Zn, Pb, Cu and Cr were much less than the recommended limits (**Table 2**, I-11) [30].

Tzanakos et al. investigated solidification of bottom ash {27.77, 39.74, 5.16} and fly ah {89.20, 6.00, 0.00} generated from incinerated medical waste in MK-based GP using sodium hydroxide and silicate activators. The TCLP leaching test indicated that the concentrations of Cr, Ni, Zn, Cd, Ba and Pb in the leachate were lower than the permitted limits and lower than the amounts leached out of unreacted bottom and fly ash (**Table 2**, I-12). The mixing of bottom, fly ash and MK with $CaCO_3$ was shown to improve the mechanical strength of the produced GP without affecting the leachability of heavy metals [7]. Thus, both the works of Guo et al. [30] and Tzanakos [7] reflects the positive effect Ca on immobilization of heavy metals in GP matrix.

3.1.4. Exhausted adsorbents

Xu et al. reported an interesting study on solidification of zeolite A {0.54, 54.6, 27.8} loaded with ^{90}Sr radionuclide in MK {0.13, 45.25, 43.92}-based GP using sodium silicate as activator. The performance of solidification in GP was compared with that in OPC. The mechanical strength of the produced GP was about 40 MPa compared with 11 MPa in the case of OPC. Furthermore, the GP was found to have better leaching resistance than OPC in different leaching solutions: deionized water, H_2SO_4 solution at pH 1, 4% (w/w) $MgSO_4$ solution and acetic acid solution at pH 3.6. This study proved that solidification in GP is complementary to adsorption process where the exhausted adsorbents containing pollutants could be solidified in a form with high mechanical strength and low leachability for construction purpose or landfill [31].

3.1.5. Nuclear waste

The solidification of nuclear waste by GP was reviewed by Vance and Perera and will not be discussed here. It was concluded that GP can have significant advantage over other cements: fire resistance, good immobilization and lake of freeze-thaw problem. There has been actual disposal of low level nuclear waste in GP on an industrial scale in the Slovak Republic [11].

3.2. Inter-solidification of pollutants in geopolymer

The present section reviews literature on immobilization of highly soluble metal salts ($Cu(NO_3)_2$, $Pb(NO_3)_2$, $Cd(NO_3)_2$, etc.) in aluminosilicate-rich sources like FA (type F), MK and BFS, etc. The heavy metal nitrates are usually introduced during the synthesis of GP and can be added either to the aqueous activator (sodium hydroxide and sodium silicate solution) or added to solid aluminosilicate precursor (**Figure 1**). This technique can be applied for solidification of waste water contaminated with heavy metals. As will be discussed in the forthcoming paragraphs, the effect of incorporation of heavy metal nitrates on the mechanical strength of the produced GP and the % leaching of heavy metals from the GP matrix are the two major criteria for evaluating the efficiency of solidification. Other microstructural properties of GP (revealed by XRD, SEM and FTIR) are not usually affected by incorporation of heavy metals.

3.2.1. *Alkali and alkaline earth metals*

Lee et al. investigated the effect of incorporation of inorganic salts (0.18 mol/kg FA) KCl, $CaCl_2$, $MgCl_2$, K_2CO_3, $CaCO_3$, $MgCO_3$, $Ca(OH)_2$ and $Mg(OH)_2$ on the mechanical strength of GP prepared from type F FA {3.5, 50.0, 28.0} and kaolin {0.2, 54.5, 29.4}. Two systems of alkali activators were employed: the first contained 15 M NaOH and the second contained 20 M NaOH in addition to sodium silicate. The first system was found to have higher mechanical strength. $Ca(OH)_2$ and $Mg(OH)_2$ did not exert any negative effect on the mechanical strength, KCl, $CaCl_2$, $MgCl_2$ decreased the mechanical strength and K_2CO_3, $CaCO_3$, $MgCO_3$ increased the mechanical strength of GP. The conclusion was that chloride contamination and excessive alkalinity in the activating solution should be avoided in GP synthesis [32].

Li et al. demonstrated that FA {63.86, 21.87, 4.33}-based GP are more effective hosts for $^{133}Cs(I)$ than ordinary Portland cement (OPC). $^{133}Cs(I)$ was incorporated (2% w/w) into GP and OPC, separately. The concentration of Cs(I) leached out of the GP was found to be 5.4 and 9% of the leaching concentration of OPC when deionized water and $MgSO_4$ solutions were used as leaching reagents, respectively. This was ascribed to the fact that GP has much lower porosity than OPC. On the other hand, when H_2SO_4 was used as leaching reagent, Cs(I) leaching from GP was comparable to leaching from OPC. The high extent of leaching in H_2SO_4 leaching reagent (30 times that of deionized water) was ascribed to dissolution of $Ca(OH)_2$ from GP and OPC. The GP-containing Cs(I) was superior to OPC in many properties like higher mechanical strength, thermal stability and acid resistance. The mechanical strength of GP was 57.15 MPa compared with 33.73 in the case of OPC. The GP block can retain a compressive strength of 30 MPa after 2 h calcination at 1000°C. The mass loss of GP after 60 days in acetic acid/acetate buffer (pH 3.6) was 7.1% that of OPC [33].

3.2.2. *Pb(II), Cu(II), Cd(II), Cr(III)*

Van Jaarsveld et al. studied solidification of Cu(II) and Pb(II) nitrates (0.1% w/w) in GP prepared from FA {8.2, 50.1, 28.3} and either MK or kaolin additives and using sodium hydroxide activator. The concentrations of Cu and Pb in the TCLP leaching solutions were higher than the allowed limit (**Table 3**, I-1). Furthermore, the % leaching was relatively high (23–65%) [34]. The compressive strength of Pb-containing GP samples (33.7 MPa) was found to be slightly higher than Cu-containing GP samples (28.1 MPa). Two basic mechanisms for immobilization of heavy metals in GP matrix were proposed. The first is physical: positively charged metal Pb(II) or Cu(II) ions balance the negatively charged Al tetrahedra of GP. The second is chemical: metal ions are bonded to the silicate chain through oxide and/or hydroxide links [35].

Phair et al. incorporated 0.5% w/w Pb(II) and Cu(II) nitrates in FA-based GP. The FA {6.1, 48.5, 29.6} was either used alone or mixed with kaolin {0.1, 52.4, 28.6}, MK or K-feldespar {0.21, 67.1, 17.6} additives using sodium hydroxide and/or silicate activators. The efficiency of immobilization of Pb(II) followed the order: FA-GP > FA/kaolin-GP > FA/feldspar-GP > FA/MK-GP as indicated by TCLP leaching tests (**Table 3**, I-2). The concentrations of heavy metals were higher than the TCPL limit and the % leaching were relatively high. There was no correlation between

GP matrix	Code	Leaching test	Mass % of metal solidified	Cu(II)	Pb(II)	Cr(VI)	Cd(II)	Cr (III)	Ni (II)	Ref.
FA/MK	I-1	TCLP, pH 3, 24 h	0.1	22 (55)	23 (58)					[34]
FA/kaolin				26.1 (65)	9.1 (23)					
FA/kaolin	I-2	TCLP, 28 days	0.5	85.7 (43)	8.5 (43)					[36]
FA/MK				101 (51)	17.5 (9)					
FA/feldspar				116 (58)	11 (6)					
FA				113 (57)	7.5 (4)					
MK/FA	I-3	TCLP, pH 3, 24 h	0.1	17 (125)	17 (68)					[37]
		HCl solution, pH 3		8 (59)	3 (12)					
Kaolin/FA		TCLP, pH 3, 24 h		9 (67)	7 (28)					
		HCl solution, pH 3		17 (125)	10 (40)					
FA/MK	I-4	TCLP	0.1	0.0075	0.0025		0.020	0.001		[38]
BFS/MK(1:1)	I-7	TCLP, 24 h	0.1	0.176 (0.44)	0.292 (0.73)					[42]
			0.2	0.384 (0.48)	0.864 (0.72)					
			0.3	0.396 (0.33)	0.192 (0.16)					
FA	I-5	H_2SO_4, pH 1, 90 days	0.5		(0.4)	(88)	(37)			[15]
		5% $MgSO_4$, 90 days			(0.0)	(75)	(0.02)			
		5% Na_2CO_3, 90 days			(0.1)	(78)				
		Water, 90 days			(0.004)	(80)	(0.04)			
MK, sodium silicate activator	I-8	Deionized water, L/S ratio 6	0.1	22.1 (28.4)	10.3 (4.10)		20.5 (14.82)			[12]
		0.1 M HCl		15.0 (21.04)	7.5 (2.71)		10.7 (7.28)			
		0.1 M NaCl		11.3 (15.51)	4.7 (1.81)		7.7 (4.89)			
		1.0 M NaCl		7.4 (0.82)	18.1 (7.71)		10.0 (6.16)			
		0.1 M NaOH		14.3 (21.22)	7.1 (2.52)		7.7 (5.01)			
Ferronickel slag	I-9	TCLP, pH 3 24 h	0.5 nitrates	25.5 (4)	7.9 (0.5)			0.3	34.7	[44]
			0.5 sulfate	46.6 (7)	9.0 (0.5)			0.1	47.3	
FA	I-6	EN-12457-2 deionized water, 28 days	0		1.0					[40]
			0.5		4.0 (0.8)					
			1.0		7.0 (1.4)					

GP matrix	Code	Leaching test	Mass % of metal solidified	Cu(II)	Pb(II)	Cr(VI)	Cd(II)	Cr (III)	Ni (II)	Ref.
Mechanically activated FA			0		0.5					
			0.5		0.9 (0.2)					
			1.0		1.0 (0.2)					
FA	I-10		0.5			126 (25)				[46]
			1.0			338 (34)				
			2.0			735 (37)				
Mechanically activated FA			0.5			44 (9)				
			1.0			219 (22)				
			2.0			530 (27)				
USEPA TCLP allowed limit (mg/L)					5	5	1	5		

MK: metakaolin; FA: fly ash; BFS: blast furnace slag. L/S ratio: liquid leaching solution (mL) to solid geopolymer (g).

Table 3. Leaching of heavy metals (mg/L) from geopolymer matrix containing solidified heavy metal nitrates—intersolidification (values in brackets give calculated % leaching).

the efficiency of immobilization and the compressive strength where the order of mechanical strength was FA/kaolin-GP > FA/MK-GP > FA/feldspar-GP > FA-GP

	FA-GP	FA/kaolin-GP	FA/feldspar-GP	FA/MK-GP
% leaching of Pb	7.5	8.5	11	17.5
Mechanical strength (MPa)	7.7	32.7	13.9	26.8

The efficiency of immobilization of Cu was lower than that of Pb. An interesting conclusion withdrawn from this study was that Cu and Pb are less labile to leaching from bulky poly-silicates of GP than from their mono-silicates and hydroxides [36]. In a similar study, van Jaarsveld et al. compared the effect of addition of 15% (w/w) kaolin or MK to FA-based GP on solidification of 0.1% (w/w) $Pb(NO_3)_2$ and $Cu(NO_3)_2$. The FA used was of type F {8.2, 50.1, 28.3), sodium hydroxide was used as activator and the Si/Al molar ratio was 1.5. The results are given in **Table 3**, I-3. MK was shown to be preferred over kaolin as additive to FA when high resistance to HCl leaching solution (pH 3.3) is required [37].

Xu et al. solidified Pb(II), Cd(II), Cu(II), and Cr(III) nitrates in GP prepared from FA {1.88, 47.42, 30.90} and MK {0.13, 53.3, 41.8} additive using potassium hydroxide and silicate activators. TCLP leaching tests (**Table 3**, I-4) indicated negligible leaching concentrations of heavy metals ≤0.02 mg/L (% leaching <1%) which are much lower than the TCLP limits. Except for Cr

(III), It has been found that heavy metals improve the compressive strength of GP. Furthermore, immobilization of Pb(II) and Cr(III) were better than Cu(II) and Cd(II). Increasing the % sodium hydroxide from 2 to 6% in GP mix design was found to result in decreasing of the concentration of Cu(II) and Cd(II) in the leaching solution [38]. The high efficiency of immobilization of heavy metals compared the work of Van Jaarsveld [34, 35, 37] may be due to the use of two sodium silicate in addition to sodium hydroxide.

Zhang et al. investigated solidification of 0.5% (w/w) Pb(II) and Cd(II) nitrates in FA {5.1, 46.4, 28.3}-based GP prepared using sodium hydroxide and silicate activators. H_2SO_4 (pH 1), 5% (w/w) $MgSO_4$ and 5% Na_2CO_3 leaching solutions were used. H_2SO_4 leaching solution showed the highest % leaching. Pb(II) resist leaching using all these leaching solution and Cd(II) also showed similar behavior except in the case of H_2SO_4 leaching solution (**Table 3**, I-5). The mechanism of immobilization suggested was that heavy metal ions undergo chemical bonding either with the GP gel or with the low solubility silicate or aluminate phases. Another possible mechanism was that the soluble salts of heavy metals may undergo some chemical conversion in the highly alkaline conditions of GP synthesis and then trapped in the GP matrix [15].

Guo et al. studied solidification of 0.025% (w/w) $Pb(NO_3)_2$ in FA {20.0, 38.0, 19.0}-based GP using sodium hydroxide and silicate activators (mass ratio, 5 NaOH:35 sodium silicate:60 FA). Incorporation of Pb induced slight reduction in compressive strength. The leaching concentration of Pb was 0.003 mg/L and the % leaching was 0.024%. The shift of FTIR OH stretching to lower frequency in the case of GP-containing heavy metal compared with GP-blank [39] suggests binding of nonbridging Si—OH and Al—OH to heavy metal.

Nikolic et al. studied the effect of mechanical activation of FA {5.67, 62.13, 17.20} on the efficiency of immobilization of Pb(II) nitrate in FA-based GP. Mechanical activation involves milling FA before reacting with sodium hydroxide and silicate activators. EN leaching test indicated that Pb (II) was more effectively immobilized in mechanically activated FA-based GP (the leaching concentration is less than the allowed limit) than untreated FA-based GP (**Table 3**, I-6). This was ascribed to the fact that mechanical activation of FA reduces the porosity, increase the compactness (as reflected by SEM) and increase the mechanical strength of the GP [40].

Guo et al. investigated solidification of up to 8% (w/w) of three Pb compounds (PbO, $PbSO_4$ and PbS) in FA {7.6, 52.1, 23.5}-based GP using sodium hydroxide and silicate activators (Si/Al molar ratio 2.3). The three compounds have variable solubility: PbO is soluble in acetic acid and NaOH, $PbSO_4$ is insoluble in acetic acid and soluble in NaOH, and PbS is insoluble in both solutions. Incorporation of PbO was found to increase the mechanical strength of GP while $PbSO_4$ and PbS have the reverse effect. The % leaching was very low (<1%), but the leached Pb concentration (6–27 mg/L) was higher than the TCLP limit [41].

MK and BFS-based GP received less attention than FA-based GP as hosts for highly soluble heavy metal salts. However, some studies will be discussed here. Yunsheng et al. investigated solidification of Pb(II) and Cu(II) in GP prepared from (1:1, w:w) MK {0.6, 62.97, 26.91) and BFS {41.7, 34.20, 14.20} using sodium hydroxide and silicate activators. TCLP tests (**Table 3**, I-7) indicated that MK/BFS-based GP can very effectively immobilize Cu (II) and Pb(II) with % leaching less than 1% and concentrations of heavy metals in leaching solution less than the

TCLP limit. Kinetics of leaching revealed that the heavy metals rapidly leached out the GP and reach steady concentration after 4 h. The mechanical strength of the produced GP was high and slight reduction in mechanical strength (from 72.68 to 70.28 MPa) was observed when the heavy metal dosage increases from 0.1 to 0.3% [42]. Perera et al. showed that heating of MK-based GP containing 1% (w/w) $Pb(NO_3)_2$ from 200 to 800°C, resulted in an increase of the leached Pb concentration from 2.8 to 14.8 mg/L [43]. El-Eswed et al. studied immobilization of Pb(II), Cd(II) and Cu(II) in MK {0.4, 50.62, 46.21}-based GP using sodium silicate activator either alone (Na/Al molar ratio 1) or mixed with sodium hydroxide (Na/Al molar ratio 2). The Si/Al molar ratio used was 2 for both cases. In most cases, one and two activators have similar immobilization performance which suggests that the high alkalinity is not the limiting factor for immobilization of heavy metals. The % leaching and the concentrations in the leaching solutions are given in **Table 3**, I-8. The concentrations of Pb and Cd were higher than the allowed limit. When water was used as leaching solution, the efficiency of immobilization was Pb(II) > Cd(II) > Cu(II) which is consistent with the order of decreasing ionic radius. An interesting observation was that the efficiency of immobilization of total ions (revealed by conductivity measurements) increases with increase in efficiency of immobilization of heavy metals, which excluded the mechanism those heavy metals cations were exchanged with the alkali metals in the Al tetrahedra. Evidences for the presence of nonbridging Si—OH and Al—OH in the GP were obtained from infrared study which may be the active sites for complexation with heavy metal ions [12].

Komnitsas et al. investigated solidification of (0.5% w/w) Pb(II), Cu(II) Cr(III) and Ni(II) nitrates and sulfates in ferronickel slag {3.73, 32.74, 8.32} using potassium hydroxide and sodium silicate activators (mass ratio, 82 slag:6H_2O:3KOH:9 sodium silicate). The results (**Table 3**, I-9) reflected that the efficiency of immobilization was very high >90%. The GP undergo 70% reduction in mechanical strength when 3% $PbSO_4$ was incorporated. Except in the case of Cr(III), the metal concentration leached in the case heavy metal sulfates were more than nitrates [44]. El-Eswed et al. studied solidification of 200–1000 ppm solutions of Pb(II), Cu (II), Cd(II) and Cr(III) nitrates solutions by mixing with zeolite, kaolin and NaOH (1:7:7:1 mass ratio) followed by pressing under 15 MPa. The mechanical strength of the obtained GP was about 18 MPa. The % leaching of Pb ranged from 25 to 50 depending on the leaching solutions used; deionized water, 0.1 M NaCl, 1.0 M NaCl, 0.1 M HCl and 0.1 M NaOH leaching solutions. The worst immobilized heavy metals were Pb(II) and Cu(II) and the most effectively immobilized were Cd(II) and Cr(III) [45].

3.2.3. *Cr(VI) anions*

Zhang et al. investigated solidification of 0.5% Cr(VI) (Na_2CrO_4) in FA {5.1, 46.4, 28.3}-based GP prepared using sodium hydroxide and silicate activators. Cr(VI) was not effectively immobilized where the % leaching ranged from 75 to 88% [15]. Guo et al. studied solidification of 0.025% (w/w) Cr(VI) (CrO_3) in FA {20.0, 38.0, 19.0}-based GP using sodium hydroxide and silicate activators (5 NaOH:35 sodium silicate:60 FA). The leaching concentration of Cr was 0.015 mg/L and the % leaching was 12% [39]. To improve efficiency of immobilization, solidification of Cr(VI) (K_2CrO_4) in mechanically activated FA was investigated. Although mechanical activation of fly ash improved immobilization of Cr, the leaching concentration was much more than the allowed limit (**Table 3**, I-10). Thus Cr(VI) is not easy to be immobilized effectively in FA-based GP. GP has a limited capacity for heavy metals and the % leaching of Cr increases from 8.84 to 26.5% by increasing dosage of Cr from 0.5 to 2.0% (w/w) [46]. The addition of Cr to the GP matrix did not affect the mechanical strength of GP [46].

Chen et al. [47, 48] effectively solidified highly soluble $K_2Cr_2O_7$ in MK-based GP using sodium hydroxide and silicate activators by addition of Fe^{2+} reducing agent. The mass ratio of MK, sodium silicate solution and NaOH were set to be 6:5:1. The compressive strength of GP containing 0.1% (w/w) Cr(VI) was 60 MPa which decreased to 31 MPa when the amount of Cr(VI) increased to 0.8% (w/w) (keeping the Fe^{2+}/Cr(VI) molar ratio 4). The concentration of Cr leached using TCLP test decreased from 20.05 to 0.42 mg/L when the amount of $FeCl_2{\cdot}4H_2O$ increased from 0.5 to 3% (w/w) keeping the 0.1% Cr (w/w) constant. On the other hand, the concentration of Cr leached increased only from 0.0135 to 0.42 mg/L when the % Cr (VI) increased from 0.1 to 0.8% (w/w) keeping Fe^{2+}/Cr(VI) molar ratio = 4. The mechanism of immobilization suggested was that Fe^{2+} reduced Cr(VI) to Cr(III) which could attached to the negatively charged Al tetrahedra of the GP framework [47]. Chen et al. showed that the concentration of Cr leached out of GP containing 0.1% (w/w) Cr(VI) dropped from 47 to 0.69 mg/L when Fe^{2+} was added to GP during reaction of the MK with the alkali activator. However, the drop was more (0.0942 mg/L) if Fe^{2+} was added to Cr(VI) in water first [48].

3.2.4. *As(IV) anions*

Fernandez et al. studied solidification of 1% (w/w) $NaAsO_2$ in MK {0.0, 49.85, 36.34} and FA {4.39, 51.51, 27.47}-based GP, separately, using sodium hydroxide and silicate activators in the former and sodium hydroxide activator in the latter. The mechanical strength of MK-GP (4 MPa) was less than that of FA-GP (7.6 MPa). The % leaching of As from MK-GP and FA-GP was high (67.6 and 49.2%, respectively) and and the concentrations of As leached were higher than the TCLP limit. Interestingly, the low level of Si and Al leached (about 1%) showed that the GP matrix is stable in the aggressive leaching solution of acetic acid buffer at pH 3 (TCLP test) [49].

3.2.5. *Solidification of organic pollutants*

Few works were reported on solidification of organic pollutants in GP. Gokhale reported a technique for removal of phenols from aqueous solution using natural zeolite (clinoptilolite) followed by encapsulation in FA {1.7, 55.0, 28.4}-based GP. The mass ratio of precursors was 10 FA:1 kaolin:1.2 potassium hydroxide:2.2 sodium silicate:1.6 water. The mechanical strength of GP containing 1% (w/w) phenol and chlorophenol was 35 and 40 MPa, respectively. The TCLP leaching concentration of phenols was less than 2 mg/L and the corresponding % leaching about 40% [50]. Shvarzman studied solidification of 0.3% (w/w) phenol in FA {0.0, 47.04, 29.37} and MK {0.0, 52.18, 43.36}-based GPs, separately, with Si/Al molar ratio of about 3 and using sodium hydroxide and silicates activators. The compressive strength of FA-GP and MK-GP was 90 and 50 MPa, respectively, which was not influenced by addition of phenols. The concentration of phenol leached out of the FA-GP and MK-GP (using distilled water as leaching reagent, liquid/solid ratio 40 mL/g) was 12 and and 2.9 mg/L, respectively. The corresponding % leaching was 16 and 4%, respectively [51].

Cantarel et al. solidified liquid oil (simulating radioactive oil waste from nuclear industry) in MK {0.3, 54.4, 38.4}-based GP using sodium hydroxide and silicate activators with Si/Al and Na/Al molar ratios 3.8 and 1.0, respectively. The compressive strength of GP containing oil ranged from 22 to 31 MPa depending on the oil fraction (from 7 to 20% (v/v)). The % leaching was less than 0.19% using water leaching solution (the increased alkalinity during leaching was neutralized with HNO_3). The mechanism suggested for immobilization of oil was that the alkali from the aqueous activator reacts with alkanoic acids in the oil producing surfactants which reduces the interfacial tension between the oil and the aqueous phase [52].

3.3. Adsorption of pollutants onto geopolymer

The chemical structure of GP is composed of a negatively charged aluminosilicate framework balanced by alkali metals (such as Na^+ and K^+). The latter can be exchanged with heavy metals cations in aqueous solution and thus GP could be used for the removal of heavy metal cations from water. It is worth to mention that in all the reviewed studies below, the prepared GP adsorbent was extensively washed with distilled water (until neutral pH 7 is achieved) to remove excess alkali used in preparation of GP. This was an essential step to avoid precipitation of heavy metal hydroxides which result in overestimation of the calculated adsorption capacity of GP.

3.3.1. Fly ash-based geopolymeric adsorbents

Many studies indicated that FA-based GP is a potential adsorbent for heavy metals ions. Some results of these studies are summarized in **Table 4**. Some of the values of adsorption capacity (Q_m, mg/g)) were 99 (Cu(II)) [53], 134.95 Pb(II) [54] and 98.84 Cu(II) [55], 69.85 (Cu(II) [56] and 50.03 (Co(II)) [57]. The affinity constants (K_L, L/mg) were 0.0607 (Pb(II)) [54] and 0.061 (Cu(II)) [55]. Interestingly, the adsorption capacity of Cu(II) on the different FA-based GPs reviewed in this section (**Table 4**) is 98 ± 12 and the affinity constant K_L is 0.08 ± 0.03 which reveals that the values of adsorption parameters is somewhat independent on the FA source, GP preparation conditions and adsorption conditions. The adsorption rate constants (k_2, mg/g min) were of the same order; 1.8×10^{-2} (Cu(II)) [55], 2.8×10^{-2}–1.5×10^{-2} (Cu(II)) [56], 5.4×10^{-2}–4.2×10^{-2} (Co (II)) [57]. However, the adsorption process was reported sometimes to be slow (30 h equilibration time) [53] and sometimes fast (30 min [54] and 15 min) [55]. The adsorption capacity Q_m obtained for GP was much higher than the fly ash itself (0.1 mg/g), which revealed the efficiency of geopolymerization in creating new sites for adsorption. Furthermore, geopolymerization increases the surface area from 8.4 m^2/g in FA to 56.0 m^2/g in GP [53].

The work of Muzek was distinguished from others in studying the amounts of Na, K, Ca, Mg, Al and Si leached out of the GP as a result of adsorption of Cu(II) and Co(II). The total amount of ingoing heavy metals was found to be higher than the amount of Na^+ outgoing, indicating that the adsorption of Cu(II) and Co(II) is not only ion exchange with Na^+ (balancing the negatively charged Al tetrahedra). The amounts of Si and Al leached (<3%) were small because of the stability of GP [58].

Adsorbent	Precursors {% CaO, %SiO_2, % Al_2O_3}	Alkali activator	Adsorption conditions	Parameter	Cu(II)	Pb(II)	Cs(I)	Ref.
FA-GP	FA {1.6, 55.0, 29.0}	Sodium hydroxide	C_i: 100–250 mg/L, pH 7, solid/liquid = 0.15 g/L	Q_m	99			[53]
				K_L	0.13			
				k_2	2.8×10^{-5}			
FA-GP	FA {1.75, 50.73, 28.87}	Sodium hydroxide	C_i: 10–140 mg/L, pH 5–6, solid/liquid = 1.4–2.0/L	Q_m	96.84	134.95		[54]
				K_L	0.061	0.0607		
				k_2	1.8×10^{-2}			
FA	FA {5.14, 29.12, 51.39}	Sodium hydroxide and sodium silicate	C_i: 100–1000 mg/L, pH 3, solid/liquid = 4.0 g/L	Q_m		53.47		[59]
				K_L		0.1769		
				k_2		3.4×10^{-4}		
FA-GP				Q_m		111.1		
				K_L		0.6429		
				k_2		2.7×10^{-3}		
Faujasite from FA-GP				Q_m		142.86		
				K_L		0.2966		
				k_2		1.3×10^{-3}		
Mesoporous FA/BFS-GP	FA {4.79, 57.0, 21.0} BFS {47.7, 32.4, 11.5}	Sodium hydroxide and sodium silicate	C_i: 10–150 mg/L, pH 6, solid/liquid = 4.4 g/L	Q_m			15.244	[61]
				K_L			0.1816	
				k_2			3.03×10^{-4}	
Porous FA/iron ore tailing- GP	FA {5.21, 29.47, 51.72} Iron ore tailing {7.63, 34.72, 16.22}	Sodium silicate	C_i: 100–200 mg/L, pH 6, solid/liquid = 3.0/L	Q_m	113.41			[62]
				K_L	0.073			
			C_i: 100–200 mg/L, pH 5, solid/liquid = 3.0/L	Q_m	100.81			
				K_L	0.069			
			C_i: 100–200 mg/L, pH 4, solid/liquid = 3.0/L	Q_m	79.31			
				K_L	0.064			

MK: metakaolin; FA: fly ash; BA: bottom ash; BFS: blast furnace slag.

Table 4. Langmuir adsorption capacity (Q_m, mg/g) and affinity constant (K_L, L/mg) and pseudo second order rate constant (k_2, g/mg min) parameters for adsorption of heavy metal on FA-based GP.

Muzek et al. found that the adsorption behavior of FA-based GP was similar to zeolite NaX [57]. Furthermore, Liu et al. showed that FA-based GP is similar in adsorption properties to faujasite zeolite. The GP was prepared first and then turned into faujasite via in situ

	Faujasite	GP	FA
Specific surface area (m^2/g)	174.35	20.48	16.45
Q_m (mg Pb(II)/g)	142.86	111.11	53.47
K_L (L/mg)	0.2966	0.6429	0.1769
k_2	13.5×10^{-4}	26.73×10^{-4}	3.418×10^{-4}

hydrothermal method (by soaking in 1.0 M NaOH solution at 70°C). Interesting trends about specific surface area and adsorption capacity were obtained (**Table 4**).

The authors concluded that both GP and faujasite have the same adsorption mechanism because they have similar adsorption behavior. However, GP was distinguished in that it has the highest K_L (affinity constant) and k_2 (rate constant) [59].

In order to improve the adsorption behavior further, Novais et al. prepared a porous GP using FA {36.72, 25.34, 6.05} and MK {0.10, 54.4, 39.40} as precursors (1:2, w:w), sodium hydroxide as activator and different amounts of H_2O_2 foaming agent. In this basic medium, H_2O_2 is decomposed to water and oxygen gas. As the amount of H_2O_2 increases from 0.30 to 1.2% (w/w), the total porosity increases from 52.0 to 78.4% and the apparent density decreases from 0.98 to 0.44 g/cm^3. However, the maximum adsorption capacity of Pb(II) was low (6.34 mg/g) and the adsorption capacity of Pb(II) was not directly related to the porosity of GP [60]. A similar study was conducted by Lee et al. for the adsorption of Cs(I) on mesoporous GP. The GP was prepared from (4:1) FA {4.79, 57.0, 21.0} and BFS {47.7, 32.4, 11.5} as precursors and sodium hydroxide and silicate as activators. The XRD of the prepared GP was found to have zeolites peaks and the cation exchange capacity was 202.04 cmol$_c$/kg which close to those of zeolites. One disadvantage was that the mesoporous GP was unstable in acidic environment at pH 2. However, the adsorption capacity of Cs(I) was low 15.244 mg/g at pH 6. The adsorption process was slow where 24 h is required to reach equilibration time, however, for pulverized samples, the equilibration time was 30 min. This was an indication that considerable time is required for diffusion of Cs(I) through the bulk of mesoporous GP. The adsorption mechanism was assumed to be ion exchange and electrostatic adsorption of Cs(I) cations on the negatively charged Al tetrahedra in the GP matrix [61].

Duan et al. studied adsorption of Cu(II) on porous GP prepared from FA {5.21, 29.47, 51.72} and iron ore tailing {7.63, 34.72, 16.22} precursors, sodium silicate activator and H_2O_2 foaming agent. The total porosity of the prepared GP ranges from 56.9 to 74.6%. The adsorption capacity of the porous GP was high (**Table 4**) and about three times that of reference (nonporous) GP [62]. As in the nonporous GP [54, 55, 59], the adsorption process was found to be endothermic and the positive value of entropy indicated that entropy is increasing by desolvation of Cu(II) ion as a result of adsorption on GP. The results for the effect of pH on the adsorption capacity Q_m are shown in **Table 4**. When the pH increases from 4 to 6, the adsorption capacity of Cu(II) Q_m increases from 79.31 to 113.41 mg/g [62]. This strong pH dependence, which was observed by other researchers [54, 59], may reveal that pH sensitive sites like Si—OH and Al—OH similar to those found in kaolin may contribute to adsorption by complexation mechanism (chemisorption) [63].

3.3.2. Metakaolin-based geopolymeric adsorbents

Cheng et al. studied adsorption of Pb(II), Cu(II), Cr(III) and Cd(II) on MK {0.2, 41.5, 19.6}-based GP synthesized using sodium hydroxide and silicate as activators. Some of the results are given in **Table 5**. The adsorption selectivity order deduced by the authors, depending on the calculated Q_m values in mg metal/g adsorbents, was

	Pb(II)>	Cd(II)>	Cu(II)>	Cr(III)
Q_m (mg metal/g adsorbent)	147.06	67.57	48.78	19.94
Hydrated ionic radius (Å)	4.01	4.26	4.19	4.61

Depending on this trend, the authors deduced that metals with large hydrated ionic radii have greater affinity for water and so they tend to remain in the aqueous phase and consequently will be weakly adsorbed on the GP. However, the correct order of selectivity should depend on Q_m values in mmol metal/g adsorbent

	Cu(II)>	Pb(II)>	Cd(II)>	Cr(III)
Q_m (mmmol metal/g adsorbent)	0.768	0.710	0.6011	0.383

This order does not support the same conclusion. The mechanism for adsorption onto GP was assumed to be ion exchange rather than specific chemical adsorption because the adsorption process was found to be endothermic. However, the desorption of Cr(III), Pb(II), Cu(II) and Cd (II) from GP loaded with maximum capacity was almost negligible (after 48 h) 14.7, 5.9, 1.9 and 2.1%, respectively, which suggests specific rather than electrostatic interaction [64].

A similar study was conducted by Lopez et al. for adsorption of Cs(I), Pb(II), Cu(II), Zn(II), Ni (II) and Cd(II) (**Table 5**) on MK {0.2, 52.0, 42.8}-based GP with variable Si/Al molar ratio. As the Si/Al ratio ranged from 1 to 5, the surface area decreased from 27.5 to 2.1 m^2/g and the bulk density increased from 0.8 to 1.23 g/cm^3. The GP with Si/Al ratio of 2 was found to have the highest adsorptivity toward Pb(II). The high selectivity for adsorption of large size atoms like Cs(I) and Pb(II) as well as the fact that adsorption process was independent on ionic strength which indicates that the adsorption process has non-electrostatic mechanism [65].

Luukkonen et al. investigated adsorption of As(III) and Sb(III) onto MK {0.06, 53.1, 40.3}-based GP and compared the results with adsorption of the same heavy metals onto BFS {38.5, 27.2, 8.4}-based GP. The specific surface area increases significantly upon geopolymerization of MK and BFS from 11.5 to 22.4 m^2/g and from 2.79 to 64.5 m^2/g, respectively. The adsorption properties are shown in **Table 5**. The adsorption capacity of the GP samples was, in general, higher than the corresponding precursors (MK and BFS), which revealed the importance of geopolymerization in generation of new active sites for metal ions. However, the adsorption capacities were low, may be because As(III) occurs primarily as arsenite and arsenate and Sb

Adsorbent	Adsorption conditions	Parameter	Pb(II)	Cu(II)	Cr(III)	Cd(II)	Ni(II)	As(III)	Sb(III)	Zn(II)	Cs(I)	Ref.
MK-GP	C_i: 50–300 mg/L, pH 4, solid/liquid = 1.5 g/L	Q_m	147.1	48.78	19.94	67.57						[64]
		K_L	0.1135	0.025	0.158	0.449						
		k_2	1.8×10^{-5}	2.3×10^{-5}	1.8×10^{-3}	8.3×10^{-5}						
MK-GP	C_i: 50–500 mg/L, pH 5, solid/liquid = 1.25 g/L	Q_m	63.40	59.22							57.81	[65]
		K_L	0.0181	0.0117							0.0211	
Porous MK-GP	C_i: 100 mg/L, pH 5, solid/liquid = 1.5 g/L	Q_m	45.1	34.5								[68]
		K_L										
		k_2		3.8×10^{-5}								
Porous MK-GP	C_i: 50 mg/L, pH 5, solid/liquid = 1.5 g/L	Q_m		52.63								[69]
		K_L		0.1359								
		k_2		2.9×10^{-3}								
MK	C_i: 3.74 mg Ni/L, 0.52 mg As/L, 0.34 mg Sb/L, pH 7–8, solid/liquid = 5.0 g/L	Q_m					0.449	0.0416	0.0102			[66]
		k_2					0.025	3.137	1.996			
MK-GP		Q_m					0.748	0.078	0.0578			
		k_2					0.119		10.944			
BFS		Q_m					0.374	0.0208	0.0034			
		k_2					0.168	0.080	3.866			
BFS-GP		Q_m					0.673	0.0104	0.0034			
		k_2					1.124	0.233	1.702			
MK-GP		Q_m					42.61			74.54		[67]
		K_L					19.6			346		
		k_2					8.4×10^{-3}			3.8×10^{-3}		

MK: metakaolin; FA: fly ash; BFS: blast furnace slag.

Table 5. Langmuir adsorption capacity (Q_m, mg/g) and affinity constant (K_L, L/mg) and pseudo second order rate constant (k_2, g/mg min) parameters for adsorption of heavy metal on MK-based GP.

(III) occurs as antimonite and antimonate (oxyanions) in aqueous solutions. The equilibration time for adsorption of heavy metals on MK and BFS GP was 24 and 6 h, respectively [66].

Kara et al. studied adsorption of Zn(II) and Ni(II) on MK {0.0, 52.9, 41.9}-based GP. The GP was prepared with SiO_2/Al_2O_3 and Na_2O/Al_2O_3 molar ratios of 3.2 and 0.7, respectively. The adsorption capacity Q_m (mg/g) of Zn(II) and Ni(II) was found to follow the order

	MK-GP>	**Zeolite>**	**Kaolinite**
Zn(II)	74.53	18.66	7.20
Ni(II)	42.61	1.98	1.69

Thus, GPs were thought to be more effective than traditional adsorbents like zeolite and kaolinite [67]. An interesting observation on this study was the high values of K_L (**Table 5**) which reflects high affinity of GP sites to interact with heavy metals.

In order to increase the surface area of MK-based GP adsorbents, Tang et al. prepared porous MK {0.0, 56.91, 42.35}-based GP in the presence of sodium dodecyl sulfate foaming agent with Si/Al and Na/Al molar ratios 1.6 and 1, respectively. The prepared GP was found to have surface area 53.95 m^2/g (which is much higher than that of ordinary MK-GP (20–30 m^2/g [65, 66]) and bulk density 0.79 g/cm^3. The adsorption data (**Table 5**) showed that the adsorption capacity of porous GP toward Cu(II) is less than that of ordinary MK-based GP, which indicated that the employment of foaming agent may decrease surface active sites on the GP. Furthermore, the adsorption process was slow where 36–50 h is required to reach equilibration time which reflects diffusion limitations [68]. Similarly, porous MK-based GP was prepared by Yuanyuan, using H_2O_2 and sodium dodecyl sulfate as foaming agents. The adsorption properties are shown in **Table 5**, where the adsorption capacity of Cu(II) on the porous GP was 52.63 mg/g, which is comparable to ordinary MK-based GP (**Table 5**). However, the adsorption process was slow and needed 36 h to reach equilibration time. Furthermore, the adsorption capacity increased significantly from 1.02 to 38.55 mg/g as the pH increased from 2 to 5 [69], which indicated that the pH dependent sites are responsible for adsorption of Cu(II).

Medpeli et al. studied adsorption of As(V) as $HAsO_4^{2-}$ oxyanions onto MK-based GP. The prepared GP was found to have surface area 75 m^2/g which increases to 298 m^2/g after impregnation with iron (hydr)oxide nanoparticles. The highest adsorption capacity obtained was 0.950 mg/g which is very low compared to the values in **Table 5**. Thus, the authors concluded that fabricating media of high specific surface area and high iron hydroxide content does not necessarily give highest adsorption capacity because much of the iron hydroxide may not be accessible for sorption due to pore clogging during impregnation [70]. However, like the work of Luukkenon et al. [66], it seems that adsorption of metals that form oxyanions like arsenic onto GP is difficult.

3.3.3. *Zeolite-based geopolymeric adsorbents*

Adsorption of Cu(II), Pb(II), Ni(II), Zn(II) and Cd(II) onto kaolin-/zeolite-based GP was investigated. The GP was made from low cost natural kaolin and zeolite (phillipsite) using sodium hydroxide as activator. The adsorption capacity of GP (61.31 mg/g for Pb(II)) was higher than that of natural kaolin (9.61 mg/g) and zeolite (40.19 mg/g) [71–73]. The rate constant of adsorption (k_2) of GP samples was found to be less than that of natural kaolin due to the kinetic limitations imposed by formation GP network [73]. Furthermore, Andrejkovicova et al. investigated adsorption of Pb(II), Zn(II), Cu(II), Cd(II) and Cr(III) onto GP made from MK {0.10, 54.39, 39.36}, zeolite-clinoptilolite {3.38, 70.61, 12.06} and sodium hydroxide and silicate as activators. The adsorption data revealed the following order of adsorption capacity Q_m [74]

	Pb(II)>	Cd(II)>	Cu(II)>	Zn(II)>	Cr(III)
Q_m (mg/g)	202.72	53.99	35.71	30.79	18.02

3.3.4. *Adsorption of organic pollutants on geopolymeric adsorbents*

Few works were reported for adsorption of organic pollutants on GP. Li et al. studied adsorption of methylene blue and crystal violet on FA {1.6, 55.0, 29.3}-based GP using sodium hydroxide as activator. The adsorption capacity (Q_m) was 32.0 and 40.8 mg/g, respectively, the affinity constant (K_L) was 105 and 4.53 L/mg, respectively, and the rate constant (k_2) was 3.76×10^{-3} and 2.29×10^{-5} g/mg min, respectively. The adsorption capacities on GP were much higher than that on unreacted FA, while the rate constants of GP were much lower than those of unreacted GP [75]. Less adsorption capacity was obtained in the case of zeolite-/kaolin-based GP which was 26 mg/g [72]. However, the effect of alkali activator on methylene blue and crystal violet should be investigated since it has been reported that these dyes are hydrolyzed in the highly basic medium [76].

3.4. Cost

Solidification technologies are attracting great interest from mining and energy industries to solve waste disposal problems. According to many authors, green chemistry geopolymerization can be applied to a variety of waste sources at low cost and low energy demand and environmental impact yielding added value products [36, 77]. However, this claim must not be taken for granted and should be analyzed in this section.

The annual production of coal FA waste was estimated to be about 500 million t from which 16% is used. The disposal of the unused FA has become a series of environmental problem [78]. Although considerable research has been published on FA-based GP technology, application of this technology is not yet widespread [4]. China and Australia have less restricted regulation about the use of alternative concrete than Europe in this regard [79]. Currently, there is a commercial GP concrete producer in Australia, namely The Zeobond Group, founded by Prof. van Deventer. One of their products is E-Crete™ made from FA (the by-product of burning coal at power stations) and BFS (the by-product of steel manufacturing). E-Crete™ is now used in footpaths, driveways, house-slabs, in-situ pours, etc. Its performance is superior to OPC in chemical, salt and fire resistance [80]. Furthermore, the CO_2 emission due to production of GPs is generally reported to be 60–90% less than OPC [4, 79].

The most critical reagent in the cost of GP preparation is the alkali activator. However, the amount of activator is usually kept minimal in the FA-GP preparation (less than 10% of the total mix design). For the above reviewed works, the cost of GP adsorbent was calculated from their mix design used in synthesis of GP adsorbent (Section 3.3). The calculated costs of MK-based GP adsorbents prepared in Refs. [64, 66, 67] are US $218/t, 205/t and 190/t, respectively. The cost reduced to US $141/t in the case FA-based GP adsorbents [59]. Thus, GP adsorbents have much lower cost than granular activated carbon and peanut hulls used in water purification which is about US $2000/t, but comparable to natural zeolite (US $120/t) and coconut shell charcoal (US $250/t) [81]. The unreacted NaOH remained after geopolymerization reaction (or Na_2CO_3 due to absorption of CO_2 from atmosphere) may be useful because it can be utilized in precipitation of high concentrations of heavy metals in water treatment.

The cost of GP prepared from industrial waste for the purpose of solidification of heavy metal pollutants already present in the waste can also be calculated from the total mix design compositions reported in the above reviewed works (intra-solidification, Section 3.1). The obtained values obtained assuming negligible cost for waste transportation are US $122/t GP in the case of coal FA [19], US $91/t GP in the case of carbon steel arc furnace dust [8], US $79–194/t in the case of municipal solid waste incineration fly ash [13, 28–30], US $176–192/t GP in the case of chromite ore processing residue [25, 26] and US $200/t GP in the case medical waste incineration fly ash [7]. However, the cost ranges from US $400 to 4000/t waste, because the % waste in the GP is low (6–30%). Thus, it may not be wise to use the prepared GP in landfill and so it will worthy to use the prepared GP for construction and insulation purposes [82].

4. Conclusions

1. The present review differentiates between three lines of research that deals with application of GP in treatment of pollutants: intra-solidification, inter-solidification and adsorption.

2. Adsorption line of research has the most consistent results. Fly ash and metakaolin-based GP adsorbents, which have much lower cost than activated carbon, seems to be very effective adsorbents for heavy metal ions like Pb(II), Cd(II), Cu(II) and Zn(II). However, they are poor adsorbents for heavy metals forming oxyanions like As(III) and Sb(III). There are strong evidences that geopolymerization reaction generates new adsorption sites for heavy metals. The adsorption performance of geopolymers is similar or better than zeolites and to some extent independent on the nature of raw material and preparation conditions of GP. Note that GP has more acid resistance than zeolites. The observations that adsorption is pH dependent and ionic strength independent and the irreversibility of adsorption process all support specific rather than electrostatic interaction between heavy metal and GP sites. Strangely, increasing the porosity of GP does not lead to improvement of its adsorption capacity.

3. Intra-solidification of waste: Despite the high mechanical strength of the produced GP-containing waste, geopolymerization reduces the leaching of heavy metals from GP matrix. The concentrations of heavy metals leached are lower than the allowed limit and the % leaching is low. A number of studies reflect the positive effect Ca on immobilization

of heavy metals present in waste in GP matrix. Pb, Cu Zn are the best immobilized, while elements forming oxyanions like Sb, Mo, V and Cr are the least immobilized in GP. There were no conclusive results on the effect of Si/Al ratio, alkali activator dose and the nature of leaching solution on the efficiency of immobilization.

4. Inter-solidification presents a greater challenge because highly soluble heavy metals salts are usually incorporated in the GP preparation. On the other hand, in intra-solidification, heavy metals may exist in less soluble oxide form. In many works reviewed, the concentrations of heavy metals in the leaching solution are higher than the allowed limit and the % leaching is high. The counter ions associated with the heavy metal in inter-solidification such as sulfate and chlorides present greater difficulties in immobilization than nitrates. Fly ash-based GP is an efficient host for heavy metals nitrates if sodium silicates are used in addition to sodium hydroxide activator. Pb(II), Cu(II), Cd(II) and Cr(III) salts are much more effectively immobilized than heavy metals salts forming oxyanions like As(IV) and Cr(VI). However, addition of reducing agents makes immobilization of Cr (VI) very effective because of reduction of oxyanion Cr(VI) to cationic Cr(III). GP has a limited capacity up to 0.5% (w/w) in the GP mix design. The leaching of heavy metals is fast and reaches steady state after 4 h.

5. Two criteria must be applied to evaluate the efficiency of immobilization. The first is the concentration of heavy metal in the leaching solution, which should be less than the allowed limit. The second is that the % leaching should be small. These two criteria may give contradicted conclusions in some cases about the efficiency of immobilization.

6. The target of adsorption should be water contaminated with relatively low heavy metals concentrations (up to 100 mg/L). On the other hand, the target of solidification is water contaminated with high heavy metals concentrations up to 10,000 mg/L (assuming 0.5% w/w heavy metal in mix design of GP and 50% water content). Adsorption techniques have some advantages over solidification, for example, the purified water is useful. However, adsorption has disadvantages such as the adsorbent loaded with heavy metals could not be safely disposed by landfill. On the other hand solidification has many advantages such as utilization of water polluted with heavy metals in making construction materials.

7. Both solidification and adsorption processes can be used in an integrative manner. After removal of pollutants by adsorption on zeolite, kaolin or any aluminosilicate material, the adsorbent material can then be introduced into solidification (geopolymerization) process for final disposal in the environment or for construction purposes.

8. The relatively high cost and the high mechanical strength of the solidified GP products containing waste necessitates that the product should be used in outdoor applications like insulation, roads, tsunami walls, etc.

Abbreviations

GP geopolymer

MK metakaolin

FA fly ash from coal (type F, low calcium)

BFS blast furnace slag from iron manufacturing

OPC ordinary Portland cement
Values between brackets after MK, FA and BFS: {% CaO, % SiO_2, % Al_2O_3}

Author details

Bassam I. El-Eswed

Address all correspondence to: bassameswed@bau.edu.jo

Zarqa College, Al-Balqa Applied University, Zarqa, Jordan

References

[1] Davidovits J. Geopolymers: Inorganic polymeric new materials. Journal of Thermal Analysis. 1991;**37**:1633-1656

[2] Ducman V, Korat L. Characterization of geopolymer fly-ash based foams obtained with the. Materials Characterization. 2016;**113**:207-213. DOI: 10.1016/j.matchar.2016.01.019

[3] Melar J, Renaudin G, Leroux F, Hardy-Dessources A, Nedelec J, Taviot-Gueho C, Petit E, Steins P, Poulesquen A, Frizon F. The porous network and its Interface inside Geopolymers as a function of alkali cation and aging. Journal of Physical Chemistry C. 2015;**119**(31):17619-17632. DOI: 10.1021/acs.jpcc.5b02340

[4] Zhang Z, Provis JL, Reid A, Wang H. Geopolymer foam concrete: An emerging material for sustainable construction. Construction and Building Materials. 2014;**56**:113-127. DOI: 10.1016/j.conbuildmat.2014.01.081

[5] Zhang F, Zhang L, Liu M, Mu C, Liang YN, Hu X. Role of alkali cation in compressive strength of metakaolin based geopolymers. Ceramics International. 2017;**43**(4):3811-3817. DOI: 10.1016/j.ceramint.2016.12.034

[6] Zhang L, Zhang F, Liu M, Hu X. Novel sustainable geopolymer based syntactic foams: An eco-friendly alternative to polymer based syntactic foams. Chemical Engineering Journal. 2017;**313**:74-82. DOI: 10.1016/j.cej.2016.12.046

[7] Tzanakos K, Mimilidou A, Anastasiadou K, Stratakis A, Gidarakos E. Solidification/stabilization of ash from medical waste incineration into geopolymers. Waste Management. 2014;**34**(10):1823-1828. DOI: 10.1016/j.wasman.2014.03.021

[8] Fernández Pereira C, Luna Y, Querol X, Antenucci D, Vale J. Waste stabilization/solidification of an electric arc furnace dust using fly ash-based geopolymers. Fuel. 2009;**88**(7): 1185-1193. DOI: 10.1016/j.fuel.2008.01.021

[9] Duxson P, Fernandez-Jimenez A, Provis JL, Lukey GC, Palomo A, Van Deventer JSJ. Geopolymer technology: The current state of the art. Journal of Materials Science. 2007;**42**:2917-2933

[10] Duxson P, Mallicoat SW, Lukey GC, Kriven WM, Van Deventer JSJ. The effect of alkali and Si/al ratio on the development of mechanical properties of metakaolin-based geopolymers. Colloids and Surfaces A: Physicochemical and Engineering Aspects. 2007;**292**:8-20. DOI: 10.1016/j.colsurfa.2006.05.044

[11] Vance ER, Perera DS. Geopolymers for nuclear waste immobilization. In: Provis JL, Van Deventer JSJ, editors. Geopolymers: Structure, Processing, Properties and Industrial Applications. Oxford: CRC Press and Woodhead Publishing Limited; 2009. pp. 401-420

[12] El-Eswed BI, Aldagag OM, Khalili FI. Efficiency and mechanism of stabilization/solidification of Pb(II), Cd(II), Cu(II), Th(IV) and U(VI) in metakaolin based geopolymers. Applied Clay Science. 2017;**140**:148-156. DOI: 10.1016/j.clay.2017.02.003

[13] Luna Galiano Y, Fernández Pereira C, Vale J. Stabilization/solidification of a municipal solid waste incineration residue using fly ash-based geopolymers. Journal of Hazardous Materials. 2011;**185**(1):373-381. DOI: 10.1016/j.jhazmat.2010.08.127

[14] Ogundiran MB, Nugteren HW, Witkamp GJ. Immobilisation of lead smelting slag within spent aluminate—Fly ash based geopolymers. Journal of Hazardous Materials. 2013;**248–249**(15):29-36. DOI: 10.1016/j.jhazmat.2012.12.040

[15] Zhang J, Provis JL, Feng D, Van Deventer JSJ. Geopolymers for immobilization of Cr^{6+}, Cd^{2+}, and Pb^{2+}. Journal of Hazardous Materials. 2008;**157**:587-598. DOI: https://doi.org/10.1016/j.jhazmat.2008.01.053

[16] Foo KY, Hameed BH. Insights into the modeling of adsorption isotherm systems. Chemical Engineering Journal. 2010;**156**(1):2-10. DOI: https://doi.org/10.1016/j.cej.2009.09.013

[17] Tan KL, Hameed BH. Insight into the adsorption kinetics models for the removal of contaminants from aqueous solutions. Journal of the Taiwan Institute of Chemical Engineers. 2017;**74**:25-48. DOI: https://doi.org/10.1016/j.jtice.2017.01.024

[18] McLellan BC, Williams RP, Lay J, van Riessen A, Corder GD. Costs and carbon emissions for geopolymer pastes in comparison to ordinary portland cement. Journal of Cleaner Production. 2011;**19**:1080-1090. DOI: https://doi.org/10.1016/j.jclepro.2011.02.010

[19] Bankowski P, Zou L, Hodges R. Reduction of metal leaching in brown coal fly ash using geopolymers. Journal of Hazardous Materials. 2004;**114**:59-67. DOI: https://doi.org/10.1016/j.jhazmat.2004.06.034

[20] Arioz E, Arioz O, Kockar M, Leaching O. Of F-type fly ash based geopolymers. Procedia Engineering. 2012;**42**:1114-1120. DOI: https://doi.org/10.1016/j.proeng.2012.07.503

[21] Luna Y, Querol X, Antenucci D, Jdid E, Pereira F, Vale J. Immobilization of a metallurgical waste using fly ash-based geopolymers. In: World of Coal Ash (WOCA); May 7–10; Kentucky, USA. 2007

[22] Perná I, Hanzlíček T. The solidification of aluminum production waste in geopolymer matrix. Journal of Cleaner Production. 2014;**84**:657-662. DOI: https://doi.org/10.1016/j.jclepro.2014.04.043

[23] Salihoglu G. Immobilization of antimony waste slag by applying geopolymerization and stabilization/solidification technologies. Journal of the Air & Waste Management Association 2014;**64**(11):1288-1298. DOI: http://dx.doi.org/10.1080/10962247.2014.943352

[24] Rao F, Liu Q. Geopolymerization and its potential application in mine. Mineral Processing and Extractive Metallurgy Review. 2015;**36**:399-409. DOI: 10.1080/08827508.2015.1055625

[25] Sun T, Chen J, Lei X, Zhou C. Detoxification and immobilization of chromite ore processing residue with metakaolin-based geopolymer. Journal of Environmental Chemical Engineering. 2014;**2**(1):304-309. DOI: https://doi.org/10.1016/j.jece.2013.12.022

[26] Huang X, Huang T, Li S, Muhammad F, Xu G, Zhao Z, Yu L, Yan Y, Li D, Jiao B. Immobilization of chromite ore processing residue with alkali-activated blast furnace slag-based geopolymer. Ceramics International. 2016;**42**(8):9538-9549. DOI: https://doi.org/10.1016/j.ceramint.2016.03.033

[27] Antunes Boca Santa RA, Soares C, Gracher Riella H. Geopolymers with a high percentage of bottom ash for solidification/immobilization of different toxic metals. Journal of Hazardous Materials. 2016;**218**(15):145-153. DOI: https://doi.org/10.1016/j.jhazmat.2016.06.059

[28] Lancellotti I, Kamseu E, Michelazzi M, Barbieri L, Corradi A, Leonelli C. Chemical stability of geopolymers containing municipal solid waste incinerator fly ash. Waste Management. 2010;**30**(4):673-679. DOI: https://doi.org/10.1016/j.wasman.2009.09.032

[29] Zheng L, Wang W, Shi Y. The effects of alkaline dosage and Si/Al ratio on the immobilization of heavy metals in municipal solid waste incineration fly ash-based geopolymer. Chemosphere. 2010;**79**(6):665-671. DOI: https://doi.org/10.1016/j.chemosphere.2010.02.018

[30] Guo X, Hu W, Shi H. Microstructure and self-solidification/stabilization (S/S) of heavy metals of nano-modified CFA–MSWIFA composite geopolymers. Construction and Building Materials. 2014;**56**(15):81-86. DOI: https://doi.org/10.1016/j.conbuildmat.2014.01.062

[31] Xu Z, Jiang Z, Wu D, Peng X, Xu Y, Li N, Qi Y, Li P. Immobilization of strontium-loaded zeolite A by metakaolin based-geopolymer. Ceramics International. 2017;**43**(5):4434-4439. DOI: https://doi.org/10.1016/j.ceramint.2016.12.092

[32] Lee WKW, van Deventer JSJ. The effects of inorganic salt contamination on the strength and durability of geopolymers. Colloids and Surfaces A: Physicochemical and Engineering Aspects. 2002;**211**:115-126. DOI: https://doi.org/10.1016/S0927-7757(02)00239-X

[33] Li Q, Sun Z, Tao D, Xu Y, Li P, Cui H, Zhai J. Immobilization of simulated radionuclide $^{133}Cs^{+}$ by fly ash-based geopolymer. Journal of Hazardous Materials. 2013;**262**:325-331. DOI: https://doi.org/10.1016/j.jhazmat.2013.08.049

[34] Van Jaarsveld JGS, Van Deventer JSJ, Schwartzman A. The potential use of geopolymeric materials to immobilise toxic metals: Part II. Material and leaching characteristics. Minerals Engineering. 1999;**12**(1):75-91. DOI: https://doi.org/10.1016/S0892-6875(98)00121-6

[35] Van Jaarsveld JGS, Van Deventer JSJ. The effect of metal contaminants on the formation and properties of waste-based geopolymers. Cement and Concrete Research. 1999;**29**: 1189-1200. DOI: https://doi.org/10.1016/S0008-8846(99)00032-0

[36] Phair JW, Van Deventer JSJ, Smith JD. Effect of Al source and alkali activation on Pb and Cu immobilisation in fly-ash based "geopolymers". Applied Geochemistry. 2004;**19**(3): 423-434. DOI: https://doi.org/10.1016/S0883-2927(03)00151-3

[37] Van Jaarsveld JGS, Van Deventer JSJ. A comparative study of kaolinite versus metakaolinite in fly ash based geopolymers containing immobilized metals. Chemical Engineering Communications. 2004;**191**:531-549. DOI: http://dx.doi.org/10.1080/00986440490277974

[38] Xu JZ, Zhou YL, Chang Q, Qu HQ. Study on the factors of affecting the immobilization of heavy metals in fly ash-based geopolymers. Materials Letters. 2006;**60**(6):820-822. DOI: https://doi.org/10.1016/j.matlet.2005.10.019

[39] Guo X, Shi H. Self-solidification/stabilization of heavy metal wastes of class C fly ash–based geopolymers. Journal of Materials in Civil Engineering. 2013;**25**(4):491-496. DOI: https://doi.org/10.1061/(ASCE)MT.1943-5533.0000595

[40] Nikolić V, Komljenović M, Marjanović N, Baščarević Z, Petrović R. Lead immobilization by geopolymers based on mechanically activated fly ash. Ceramics International. 2014;**40** (6):8479-8488. DOI: https://doi.org/10.1016/j.ceramint.2014.01.059

[41] Guo B, Pan D, Liu B, Volinsky AA, Fincan M, Du J, Zhang S. Immobilization mechanism of Pb in fly ash-based geopolymer. Construction and Building Materials. 2017;**134**:123-130. DOI: https://doi.org/10.1016/j.conbuildmat.2016.12.139

[42] Yunsheng Z, Wei S, Qianli C, Lin C. Synthesis and heavy metal immobilization behaviors of slag based geopolymer. Journal of Hazardous Materials. 2007;**143**:206-213. DOI: https://doi.org/10.1016/j.jhazmat.2006.09.033

[43] Perera DS, Aly Z, Vance ER, Mizumo M. Immobilization of Pb in a geopolymer matrix. Journal of the American Ceramic Society. 2005;**88**(9):2586-2588. DOI: 10.1111/j.1551-2916.2005.00438.x

[44] Komnitsas K, Zaharaki D, Bartzas G. Effect of sulphate and nitrate anions on heavy metal immobilisation in ferronickel slag geopolymers. Applied Clay Science. 2013;**73**:103-109. DOI: https://doi.org/10.1016/j.clay.2012.09.018

[45] El-Eswed BI, Yousef RI, Alshaaer M, Hamadneh I, Al-Gharabli SI, Khalili F. Stabilization/solidification of heavy metals in kaolin/zeolite based geopolymers. International Journal of Mineral Processing. 2015;**137**:34-42. DOI: https://doi.org/10.1016/j.minpro.2015.03.002

[46] Nikolić V, Komljenović M, Džunuzović N, Ivanović T, Miladinović Z. Immobilization of hexavalent chromium by fly ash-based geopolymers. Composites Part B: Engineering. 2017;**112**:213-223. DOI: https://doi.org/10.1016/j.compositesb.2016.12.024

[47] Chen J, Wang Y, Zhou S, Lei X. Reduction/immobilization processes of hexavalent chromium using metakaolin-based geopolymer. Journal of Environmental Chemical Engineering. 2017;**5**(1):373-380. DOI: https://doi.org/10.1016/j.jece.2016.11.028

[48] Chena J, Wang Y, Wang H, Zhou S, Wu H. Detoxification/immobilization of hexavalent chromium using metakaolin-based geopolymer coupled with ferrous chloride. Journal of Environmental Chemical Engineering. 2016;**4**(2):2084-2089. DOI: https://doi.org/10.1016/j.jece.2016.03.038

[49] Fernandez-Jimenez A, Palomo A. Fixing arsenic in alkali-activated cementitious matrices. Journal of the American Ceramic Society. 2005;**88**(5):1122-1126. DOI: 10.1111/j.1551-2916.2005.00224.x

[50] Gokhale C. The immobilization of organic waste by geopolymerization [Thesis]. Stellenbosch: University of Stellenbosch; 2001. 189 p. Available from: http://hdl.handle.net/10019.1/52533

[51] Tavor D, Wolfson A, Shamaev A, Shvarzman A. Recycling of industrial wastewater by its immobilization in geopolymer cement. Industrial and Engineering Chemistry Research. 2007;**46**:6801-6805. DOI: 10.1021/ie0616996

[52] Cantarel V, Nouaille F, Rooses A, Lambertin D, Poulesquen A, Frizon F. Solidification/stabilisation of liquid oil waste in metakaolin-based geopolymer. Journal of Nuclear Materials. 2015;**464**:16-19. DOI: https://doi.org/10.1016/j.jnucmat.2015.04.036

[53] Wang S, Li L, Zhu ZH. Solid-state conversion of fly ash to effective adsorbents for Cu removal from wastewater. Journal of Hazardous Materials. 2007;**139**(2):254-259. DOI: https://doi.org/10.1016/j.jhazmat.2006.06.018

[54] Al-Zboon K, Al-Harahsheh MS, Bani Hani F. Fly ash-based geopolymer for Pb removal from aqueous solution. Journal of Hazardous Materials. 2011;**188**:414-421. DOI: https://doi.org/10.1016/j.jhazmat.2011.01.133

[55] Al-Harahsheh MS, Al Zboon K, Al-Makhadmeh L, Hararah M, Mahasneh M. Fly ash based geopolymer for heavy metal removal: A case study on copper removal. Journal of Environmental Chemical Engineerining. 2015;**3**(3):1669-1677. DOI: https://doi.org/10.1016/j.jece.2015.06.005

[56] Mužek MN, Svilović S, Zelić J. Fly ash-based geopolymeric adsorbent for copper ion. Desalination and Water Treatment. 2013;**52**:2519-2526. DOI: 10.1080/19443994.2013.792015

[57] Mužek MN, Svilović S, Zelić J. Kinetic studies of cobalt ion removal from aqueous solutions using fly ash-based geopolymer and zeolite NaX as sorbents. Separation Science and Technology. 2016;**51**(18):2868-2875. DOI: http://dx.doi.org/10.1080/01496395.2016.1228675

[58] Muzek MN, Svilovic S, Ugrina M, Zelic J. Removal of copper and cobalt ions by fly ash-based geopolymer from. Desalination and Water Treatment. 2016;**57**:10689-10699. DOI: 10.1080/19443994.2015.1040077

[59] Liu Y, Yan C, Zhang Z, Wang H, Zhou S, Zhou W. A comparative study on fly ash, geopolymer and faujasite block for Pb removal from aqueous solution. Fuel. 2016;**185**: 181-189. DOI: https://doi.org/10.1016/j.fuel.2016.07.116

[60] Novais RM, Buruberri LH, Seabra MP, Labrincha JA. Novel porous fly-ash containing geopolymer monoliths for lead adsorption from wastewaters. Journal of Hazardous Materials. 2016;**318**:631-640. DOI: https://doi.org/10.1016/j.jhazmat.2016.07.059

[61] Lee NK, Khalid HR, Lee HK. Adsorption characteristics of cesium onto mesoporous geopolymers containing nano-crystalline zeolites. Microporous and Mesoporous Materials. 2017;**242**:238-244. DOI: https://doi.org/10.1016/j.micromeso.2017.01.030

[62] Duan P, Yan C, Zhou W, Ren D. Development of fly ash and iron ore tailing based porous geopolymer for removal of Cu(II) from wastewater. Ceramics International. 2016;**42**(12): 13507-13518. DOI: https://doi.org/10.1016/j.ceramint.2016.05.143

[63] Gräfe M, Singh B, Balasubramanian M. Surface speciation of Cd(II) and Pb(II) on kaolinite by XAFS spectroscopy. Journal of Colloid and Interface Science. 2007;**315**(1):21-32. DOI: https://doi.org/10.1016/j.jcis.2007.05.022

[64] Cheng TW, Lee ML, Ko MS, Ueng TH, Yang SF. The heavy metal adsorption characteristics on metakaolin-based geopolymer. Applied Clay Science. 2012;**56**:90-96. DOI: https://doi.org/10.1016/j.clay.2011.11.027

[65] López FJ, Sugita S, Tagaya M, Kobayashi T. Metakaolin-based geopolymers for targeted adsorbents to heavy metal ion separation. Journal of Materials Science and Chemical Engineering. 2014;**2**:16-27. DOI: http://dx.doi.org/10.4236/msce.2014.27002

[66] Luukkonen T, Runtti H, Niskanen M, Tolonen E, Sarkkinen M, Kemppainen K, Ramo J, Lassi U. Simultaneous removal of Ni(II), As(III), and Sb(III) from spiked mine effluent with metakaolin and blast-furnace-slag geopolymers. Journal of Environmental Management. 2016;**166**:579-588. DOI: https://doi.org/10.1016/j.jenvman.2015.11.007

[67] Kara İ, Yilmazer D, Akar STI. Metakaolin based geopolymer as an effective adsorbent for adsorption of zinc(II) and nickel(II) ions from aqueous solutions. Applied Clay Science. 2017;**139**:54-63. DOI: https://doi.org/10.1016/j.clay.2017.01.008

[68] Tang Q, Ge Y, Wang K, He Y, Cui X. Preparation and characterization of porous metakaolin-based inorganic polymer spheres as an adsorbent. Materials and Design. 2015;**88**:1244-1249. DOI: https://doi.org/10.1016/j.matdes.2015.09.126

[69] Ge Y, Cui X, Kong Y, Li Z, He Y, Zhou Q. Porous geopolymeric spheres for removal of Cu (II) from aqueous solution: Synthesis and evaluation. Journal of Hazardous Materials. 2015;**283**:244-251. DOI: https://doi.org/10.1016/j.jhazmat.2014.09.038

[70] Medpelli D, Sandoval R, Sherrill L, Hristovski K, Seo D. Iron oxide-modified nanoporous geopolymers for arsenic removal from ground water. Resource-Efficient Technologies. 2015;**1**(1):19-27. DOI: https://doi.org/10.1016/j.reffit.2015.06.007

[71] Yousef RI, El-Eswed B, Alshaaer M, Khalili F, Khoury H. The influence of using Jordanian natural zeolite on the adsorption, physical, and mechanical properties of geopolymers products. Journal of Hazardous Materials. 2009;**165**:379-387. DOI: https://doi.org/10.1016/j.jhazmat.2008.10.004

[72] El- Eswed B, Yousef RI, Alshaaer M, Khalili F, Khoury H. Alkali solid-state conversion of kaolin and zeolite to effective adsorbents for removal of lead from aqueous solution. Desalination and Water Treatment. 2009;**8**:124-130. DOI: http://dx.doi.org/10.5004/dwt.2009.672

[73] El-Eswed B, Alshaaer M, Yousef RI, Hamadneh I, Khalili F. Adsorption of Cu(II), Ni(II), Zn(II), Cd(II) and Pb(II) onto kaolin/zeolite based-geopolymers. Advances in Materials Physics and Chemistry. 2012;**2**:119-125. DOI: 10.4236/ampc.2012.24B032

[74] Andrejkovičov S, Sudagar A, Rocha J, Patinha C, Hajjaji W, Ferreira da Silva E, Velosa A, Rocha F. The effect of natural zeolite on microstructure, mechanical and heavy metals adsorption properties of metakaolin based geopolymers. Applied Clay Science. 2016;**126**:141-152. DOI: https://doi.org/10.1016/j.clay.2016.03.009

[75] Li L, Wang S, Zhu Z. Geopolymeric adsorbents from fly ash for dye removal from aqueous solution. Journal of Colloid and Interface Science. 2006;**30**(1):52-59. DOI: https://doi.org/10.1016/j.jcis.2006.03.062

[76] Mills A, Hazafy D, Parkinson J, Tuttle T, Hutchings MG. Effect of alkali on methylene blue (C.I. basic blue 9) and other thiazine dyes. Dyes and Pigments. 2011;**88**(2):149-155. DOI: https://doi.org/10.1016/j.dyepig.2010.05.015

[77] Benito P, Leonelli C, Medri V, Vaccari A. Geopolymers: A new and smart way for a sustainable development. Applied Clay Science. 2013;**73**:1. DOI: https://doi.org/10.1016/j.clay.2013.03.008

[78] De Weerdt K. Geopolymers—State of the Art. Oslo: SINTEF Building and Infrastructure, Consortium Concrete Innovation Centre; 2011. 39 pp

[79] Griffin PW, Hammond GP, Norman JB. Industrial energy use and carbon emissions reduction: A UK perspective. WIREs Energy Environment. 2016;**5**:684-714. DOI: 10.1002/wene.212

[80] The Zeobond Group. E-Crete [Internet]. 2012. Available from: http://www.zeobond.com/products-e-crete.html

[81] Kurniawan TA, Chan GYS, Lo W, Babel S. Comparisons of low-cost adsorbents for treating wastewaters laden with heavy metals. The Science of the Total Environment. 2006;**366**:409-426. DOI: https://doi.org/10.1016/j.scitotenv.2005.10.001

[82] U.S. Environmental Protection Agency's Technology Innovation Office. Solidification/Stabilization Use at Superfun Sites [Internet]. 2000. Available from: https://www.epa.gov/remedytech

Chapter 6

The Application of Carbon Fiber Composites in Cryotank

Hongfei Zheng, Xuesen Zeng, Jianbao Zhang and Hongjie Sun

Additional information is available at the end of the chapter

http://dx.doi.org/10.5772/intechopen.73127

Abstract

To meet the design goal for lightweighting the next-generation launch vehicles, carbon fiber reinforced polymeric-based composites are being explored for cryogenic fuel tank applications. The applications of carbon fiber composites in liquid hydrogen (LH2) and liquid oxygen (LOX) fuel tanks were introduced in this chapter. The materials, processing, and design of DC-XA LH2 tank, X-33 LH2 tank, SLI LH2 tank, and CCTD Program tank were discussed. Lockheed Martin LOX tank and Space X LOX tank were introduced. Technology development, materials development, and development trend of cryogenic fuel tanks were discussed. Thin-ply hybrid laminates and out-of-autoclave tanks are projected for future space missions.

Keywords: carbon fiber reinforced composites, LH2 tank, LOX tank, hydrogen permeability, cryogenic properties

1. Introduction

Carbon fiber reinforced resin matrix composite materials (CFRC) are being used in the aerospace industry as a means of reducing vehicle weight. CFRC have advantages in high strength-to-weight and high stiffness-to-weight ratios. For future heavy lift launch vehicles and space exploration structures, advanced lightweight composites will be fully utilized in order to minimize vehicle weight, and CFRC in space applications requires rigorous development to demonstrate robustness, durability, and high factors of safety.

Figure 1. Delta IV heavy.

The future heavy lift launch vehicles require extremely high propellant mass fractions to achieve the designed performance. This drives the designers to incorporate lightweight materials in as many structures as possible. Propellant fuel tanks account for a large proportion in the launch vehicles, both of structural mass and geometric space. Approximately 60% of the dry mass of a launch vehicle is the fuel and oxidizer tanks. The implementation of composite cryogenic propellant fuel tanks (cryotank) for future heavy lift launch vehicles could greatly reduce the vehicle's weight by replacing the identically sized cryotanks constructed of metallic materials. United States' Committee on Materials Needs and R&D Strategy for Future Military Aerospace Propulsion Systems reported that composites offer the potential for the greatest mass reduction of all of the materials for tank [1–3].

For the case of Delta IV heavy lift launch vehicle, as shown in **Figure 1**, compared to Li-Al fuel tank, the weight saving of upperstage composite cryotanks were 43 and 26%, respectively [1].

In addition, composite design could reduce fabrication cost. Delta II faring, Delta III faring, and interstage production data have shown that composite launch vehicle structures are less expensive than metal ones [2].

Graphite-epoxy composite cryogenic tank development began at Boeing (then McDonnell Douglas) in 1987 and continues today, primarily for reusable launch vehicles (RLV) and heavy lift vehicles.

2. LH2 cryotank

The cryogenic tanks are the dominating components of the vehicle structure. To achieve weight reduction of the next-generation launch vehicles, carbon fiber reinforced polymeric-based composites are being explored for cryogenic liquid fuel tank. A composite cryotank structure can save 30% by weight than lithium aluminum alloy.

Cryogenic composite tank development began in 1987. First, most of the effort has been devoted to liquid hydrogen tanks mainly because liquid hydrogen tanks are larger than liquid oxygen tanks, lightweight materials would provide a proportionately greater weight reduction. The other reason was a common assumption that organic composite materials could not be used in LOX tanks [4].

NASA has been exploring advanced composite materials and processes to reduce the overall cost and weight of liquid hydrogen (LH2) cryotanks while maintaining the reliability of existing designs.

Composite liquid hydrogen tank development in NASA went through the National Aerospace Plane (NASP) program, the Single-Stage-to-Orbit (SSTO) vehicle X-33 program, and the recent second generation RLV applications under the Space Launch Initiative (SLI) and the Composite Cryotank Technology Demonstration Project (CCTD) (2011–2014) [5]. The fundamental issues for composite liquid hydrogen tanks are the hydrogen permeability and cryogenic mechanical properties of the composite material.

2.1. DC-XA composite LH2 cryotank

Liquid hydrogen is an essential but highly volatile fuel used in launch vehicles. It must be stored and used at −253°C, a temperature that causes most materials to become quite brittle. Liquid hydrogen also has an extremely fine molecular structure, which allows it to seep through the tiniest gaps. Therefore, liquid hydrogen tanks present an extreme challenge in engineering materials because of hydrogen permeability and cryogenic properties [2, 3].

Work in composite cryogenic tank development at McDonnell Douglas (MDA) began in 1987. MDA solved the problem of hydrogen permeation and cryogenic properties of composites successfully, and the composite LH2 tank for DC-XA was designed, developed, and fabricated. The tank was 8 feet (2.43 m) in diameter, 16 feet long (4.88 m) [6], constructed with IM7/8552 toughened epoxy material from Hercules. Automated fiber placement (AFP) was employed

Figure 2. DC-XA composite LH2 cryotank.

to manufacture the cryotank. The tank was incorporated with lightweight internal insulation modeled after the Saturn S-IVB design. The composite tank was 33% lighter than the DC-X tank.

DC-XA provided the first flight tests in real-world operating environments for composite liquid hydrogen tank in 1996, as shown in **Figure 2**. The tank passed the ground and flight tests successfully. It experienced approximately 55 pressure cycles throughout the ground and flight test program, demonstrated the acceptability of composite liquid hydrogen tank for future launch vehicle [7, 8].

2.2. X-33 composite LH2 cryotank

Lockheed Martin Space Systems Company (LM) is one of the world leaders in large cryogenic tank technology [9]. The composite LH2 cryogenic tank for X-33 vehicle was designed, developed, and built by LM and test by NASA. The tank was a conformal, load-bearing, composite sandwich structure, consisting of an outer facesheet, honeycomb core, and inner facesheet as shown in **Figure 3a**. The inner and outer facesheets were constructed with IM7/977-2 and the core was constructed with KorexTM+3-pcf honeycomb. **Figure 3b** shows the overall dimensions of the tank. The tank is 28.5 feet in length, 20.0 feet in width, and 14.0 feet in height [10, 11].

The sandwich LH2 tank of the X-33 Demonstrator (see **Figure 4**) was ground-tested at Marshall Space Flight Center on November 3, 1999. It failed this validation test when the outer skin and core of the sandwich separated from the inner skin, although the inner skin was undamaged, several fractures were observed in the outer skin in the same lobe.

After extensive study, NASA determined that the delaminations were likely caused by a combination of factors. The most probable cause of the failure was determined to be a combination of the following phenomena [1, 12–14]:

- Microcracking of the inner facesheet with gaseous hydrogen (GH2) infiltration
- Cryopumping of the exterior nitrogen (N2) purge gas

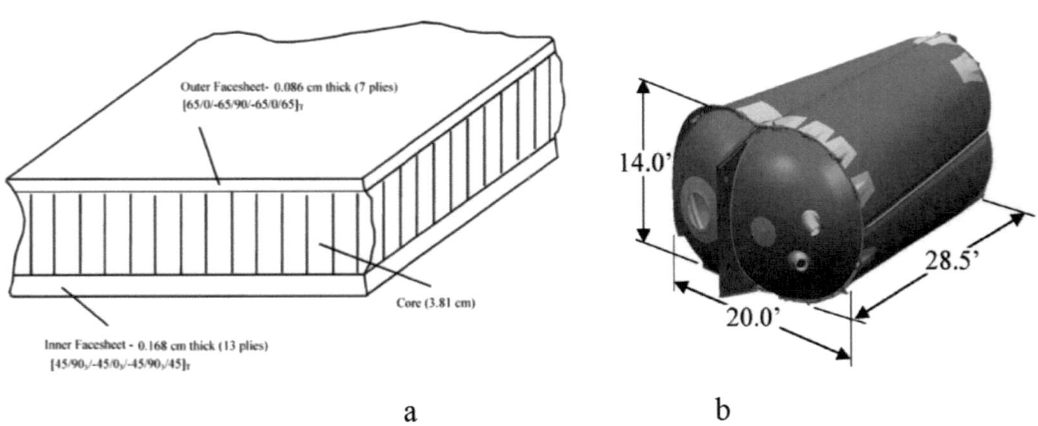

Figure 3. Lockheed X-33 composite LH2 cryotank (a) configuration of honeycomb sandwich structure and (b) tank dimensions.

Figure 4. X-33 vehicle.

- Reduced bondline strength and toughness
- Manufacturing flaws and defects
- Infiltration of GH_2 into the core, which produced higher than expected core pressures.

This setback did not discourage NASA to further develop composite cryotanks. The X-33 project was the first of its kind in large-scale composite cryotanks. The tank design was very novel with a number of cutting edge technologies. The delamination failure during the testing provided a valuable case study for NASA and Lockhed Martin to continue the composite material application in cryotank.

2.3. Space Launch Initiative (SLI) Composite Cryotank Program

Northrop Grumman designed, developed, and fabricated a cylindrical composite LH2 cryotank for Space Launch Initiative Composite Cryotank Program, as shown in **Figure 5**. The tank was constructed as a sandwich structure with carbon fiber composite skins sandwiched

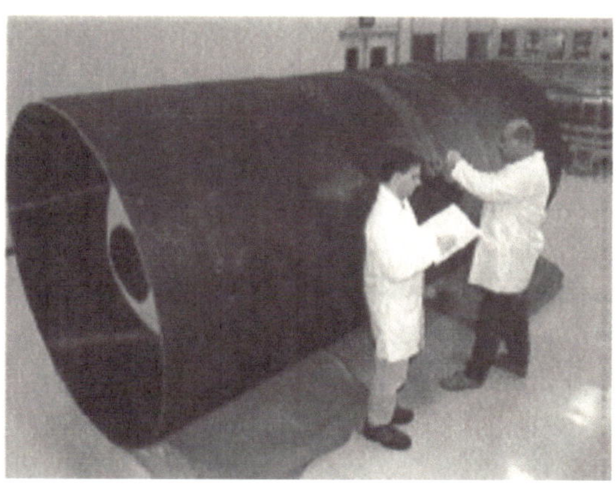

Figure 5. North Grumman's LH2 cryotank [15].

between nonmetallic honeycomb core. By using thinner carbon fiber laminars and increased number of cross laminates, the microcracks in the composite skins were significantly reduced by a factor of 16.

Other major innovations in this program included the multifunctional sandwich core and the out-of-autoclave process. Taking the advantage of the sandwich structure, hydrogen permeation was stopped by a barrier of aluminum foil between the inner skin and the honeycomb core. Any hydrogen leakage was vacuumed out through the perforated honeycomb core. In addition, the core acted as a thermal insulation layer. The cryotank manufacturing avoided the expensive autoclave by the novel ultrasonic tape lamination approach. The ultrasonic energy provided excellent compaction while depositing the carbon fiber tapes. The final composite structure was cured in an oven at ambient pressure.

The testing involved the simulated load cases close to a rocket launch. The composite cryotank, in 1.8 m diameter and 4.5 m long, was filled with liquid hydrogen. The tank was also subject to an internal pressure of 827 KPa and an axial load along the vertical axis of a launch vehicle. The composite tank was ¼ scaled size of a typical reusable launch vehicle cryotank. Further tests were performed to fill, apply internal and external loads, and drain the tank repeatedly 40 times, in order to study its structural integrity under cryogenic temperatures and its reusability [14–16].

The Northrop Grumman/NASA team proved that a cryogenic fuel tank made from composite materials has the structural integrity to withstand the mechanical and thermal stresses associated with repeated fueling and simulated launch cycles.

2.4. CCTD composite cryotank

The Composite Cryotank Technology Demonstration Project (CCTD) was part of the Space Technology Program and the Game Changing Development (GCD) Program for NASA. [17]. The program had a clear focus to deliver the proven technologies that would be deployed in future space flight demonstration. An extensive full-scale ground-based testing and lab experimentations were implemented. The development and demonstrations outlined in the CCDT Program were based on the relevant aerospace industrial experience over the last 20 years. The project successfully designed, manufactured, and tested a full-composite LH2 cryotank [18, 19]. The composite cryotank reportedly achieved 30% saving in weight and 25% in cost, compared with a baseline aluminum alloy cryotank.

NASA worked with four competing industry partners—ATK, Boeing, Lockheed Martin, and Northrop Grumman at the design phase of the project with the aim to reduce weight by 30%, reduce cost by 25%, and withstand a strain of 0.005 [20]. Four competing conceptual designs differing in materials, structures, manufacturing processes were assessed through coupon testing and finite element stress analysis.

A sample of the design requirements is shown in **Figure 6**: an inner tank that could contain pressurized LH2 at cryogenic temperatures, fitted with an outer skirt that could take the axial compression loads during launch-vehicle takeoff. All of the phase I concepts were required to use Government furnished information (GFI) IM7/977-2 lamina material property.

http://dx.doi.org/10.5772/intechopen.73127

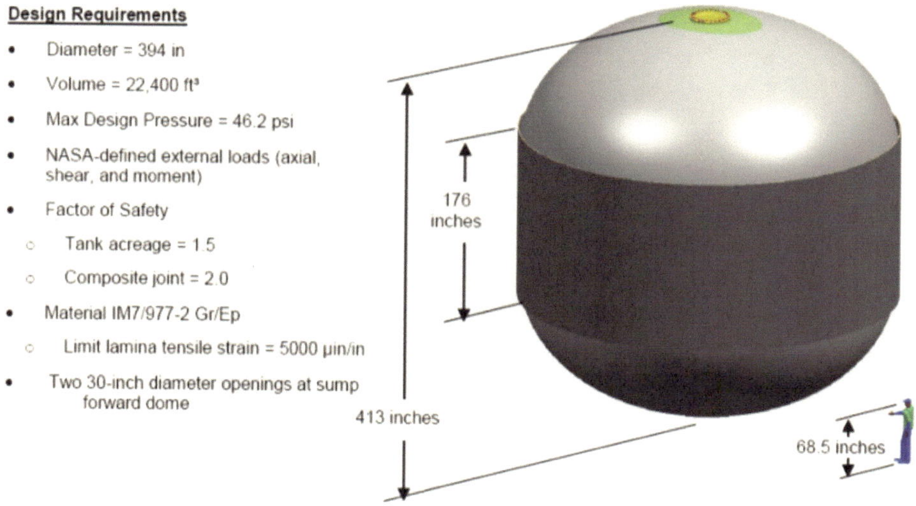

Figure 6. NASA reference Cryotank geometry and high level requirements [1].

The NASA team developed a metallic aluminum alloy cryotank concept for comparison to three industry IM7/977-2 composite concepts with the same overall dimensions: Boeing fluted core, Lockheed-Martin externally stiffened, and Northrop Grumman sandwich, as shown in **Figure 7**. The weight comparison for the four concepts is shown in **Table 1**. All three composite concepts exceeded the 30% weight reductions also had designs where the laminates were at or under the 5000 με strain limit desired by the CCTD Project when compared to the metallic cryotank. All of the three tanks were fabricated by automated fiber placement process. A cost analysis effort showed that 20–25% cost saving can be achieved by utilizing AFP process.

The results further showed that thin plies (65 or 70 g/m^2) are effective in resisting microcracks and thereby minimizing LH2 permeation.

Boeing's design and analyses showed that when designing to a 5000 με limit strain level, a 39% weight saving over a comparable aluminum-lithium tank designed using mature materials and manufacturing techniques can be realized. A cost analysis effort showed that 20–25% cost saving can be achieved by utilizing an automated fiber placement process [20].

Phase II of CCTD involved the design, analysis, fabrication, and testing of large scale (2.4-m diameter precursor and 5.5-m diameter demonstrator) composite cryotanks. The two tanks incorporated the design features and strain levels that represent a full-scale (8.4-m diameter) Space Launch System (SLS) propellant tank. Design features included a one-piece wall design that minimized tank weight.

Both tanks were fabricated at Boeing using automated fiber placement on breakdown tooling. The tanks were made of Cytec's CYCOM 5320-1 out-of-autoclave (OOA) prepreg, hybrid laminate was employed, using a combination of thick plies (145 g/m^2), which can be placed relatively quickly and enable large (up to 10-m diameter) cryotank fabrication, and thin plies (70 g/m^2), which create a microcrack-resistant laminate that helps prevent hydrogen permeation [21, 22] (**Figure 8**) .

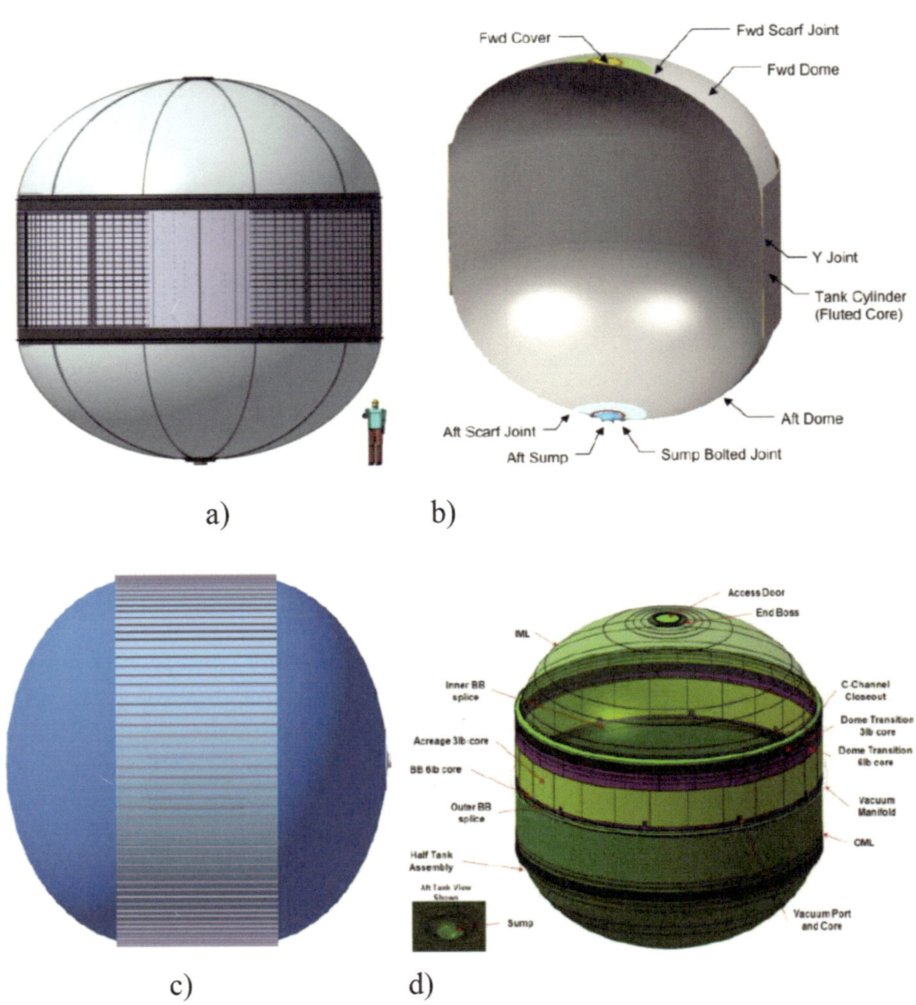

Figure 7. Four Cryotank concepts [1, 19](a) NASA al-Li concept, (b) Boeing fluted core sandwich Wall concept, (c) the Lockheed-Martin externally stiffened concept, and (d) Northrop Grumman composite honeycomb sandwich concept.

Company structural concept	Weight/lb	Weight saving/%
NASA TRL9 metallic baseline	10,925	—
Boeing fluted core	6696	38.7
Lockheed External Box Stiffened	6572	39.8
NGC Honeycomb Sandwich	6252	42.8

Table 1. Comparison of the three industry composite designs to the metallic design from NASA.

The 2.4-m diameter all-composite precursor tank was fitted with a laminate skirt that could take the axial compression loads during launch-vehicle takeoff and tested on June 25, 2013 at NASA MSFC, the tank was successfully pressure tested. The test met all requirements: stepwise fill

http://dx.doi.org/10.5772/intechopen.73127

Figure 8. 2.4-m diameter tank, a robotic arm applies composite laminate [21, 22].

Figure 9. 5.5-m diameter tank [21, 22].

with liquid hydrogen (−223°C) to 90% volume capacity followed by pressurizing the tank to 931 KPa. The 2.4 m tank was then cycled through 20 pressure/vent cycles, measuring hydrogen gas permeation on the tank dome [21] (**Figure 9**).

NASA Space Launch System (SLS) has an 8.4-m diameter (8.4 m) Ares Vupper stage. NASA chose a 5.5-m diameter test article for the second phase of the program. Boeing was the only partner to produce and test this size tank with the existing infrastructure. A 5.5-m diameter cryotank was of sufficient scale. To prevent the delaminations occurred in X-33 composite cryotanks, subscale 5.5-m diameter CCTD demonstrator tank was fitted with an innovative fluted-core skirt.

NASA completed a demanding series of tests inside the test stand at MSFC. On March 26, 2014, the 5.5 m cryotank was subjected to flight loads in combination with pressure loads. Structural loads were applied to simulate the stress during a space flight. Liquid hydrogen at 253°C was filled in the composite tank. At the same time, a cyclic pressure of 138 to 207 KPa was applied to the tank.

The 5.5 m tank was pressurized to 400 KPa, reaching a maximum acreage strain of 5136 microstrain and demonstrating safety factors above 1.5 in the scarf joints. At the 5136 test strain, the permeation performance does not meet the CCTD goals, but well within the allowable for an upper stage or boost stage composite tank application [21].

The hydrogen permeation measured in the CCTD tanks are likely due to the porosity, estimated to be approximately 3%, from the low-pressure curing having facilitated void and crack formation, even in the thin plies.

The project confirmed that composite cryotanks can achieve a 33% weight savings compared to aluminum-lithium cryotanks, and it demonstrated permeation performance that meets the allowable for upper stage and boost stage applications.

This is the first effort to successfully build and test a tank of 5.5-m diameter [21, 22].

There are several accomplishments in CCTD cryotank of phase II:

(1) 5320-1 OoA epoxy resin matrix

Boeing tested a new material-Cycom 5320-1 that does not require expensive autoclave curing. Cycom 5320-1 is a toughened epoxy prepreg resin system designed for out-of-autoclave manufacturing of primary structures. With a lower curing temperature, the resin system is suitable for prototyping where low cost tooling or vacuum-bag-only curing is required.

Cycom 5320-1 handles similarly to standard prepreg. The difference is that the vacuum-bag-only cured composites produce the quality equivalent to the autoclave process, with minimum porosity and competitive mechanical properties. It was received as thin sheets of B-stage film [23, 30].

(2) Thin ply prepreg-70 g/m^2, hybrid laminate

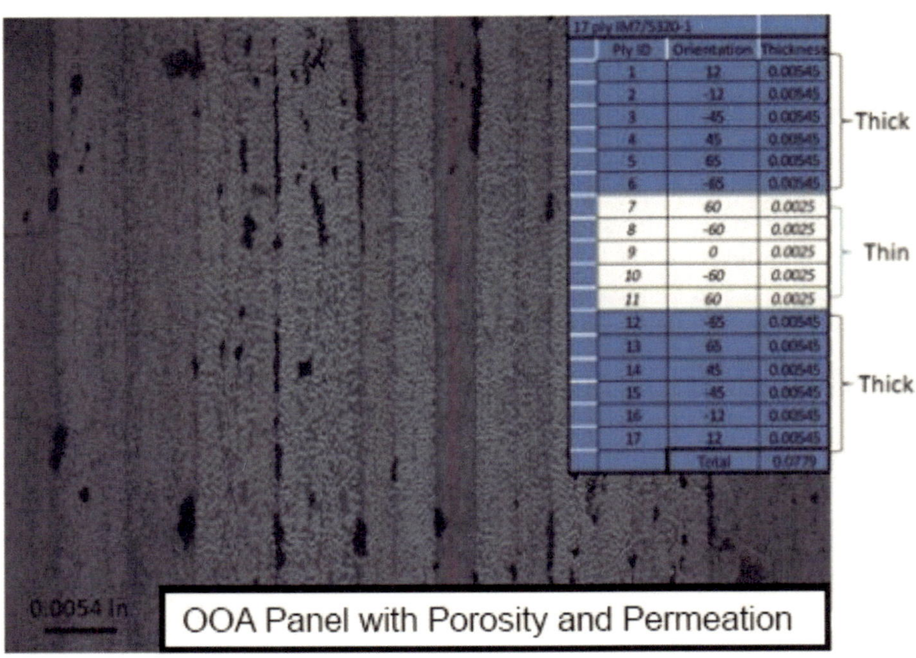

Figure 10. Thin and thick ply.

To reduce hydrogen permeation levels, hybrid laminate combination of thick plies (145 g/m^2), which can be placed relatively quickly and thin plies (70 g/m^2), which create a microcrack-resistant laminate that helps prevent hydrogen permeation was used in CCTD tanks.

For further evaluation, hybrid laminates fiber-placed panels were produced at Boeing and evaluated at MFSC. The stacking sequence of the laminate was showed in **Figure 10**. The laminates were made up of 12 plies of 5.4 mil and 5 plies of 2.5 mil material. The report noted that when thin plies are used and standard laminate consolidation is achieved, permeation performance requirements will be met with very large margins.

Thin-ply composite structures offer many advantages in composite tank manufacture. They are far more resistant to the formation of microcracks. Also, tougher resins have been developed that offer protection against microcracks (may be used in conjunction with the thin plies). The down-side of the thin plies is that they make manufacturing more difficult. Present development efforts are exploring the use of fiber placement equipment to place the thin plies. Hybrid laminates are demonstrating the same performance as the thin plies. Excellent permeability results are achieved by both methods [25].

(3) Out-of-autoclave cure processing

Figure 11. Out-of-autoclave processing.

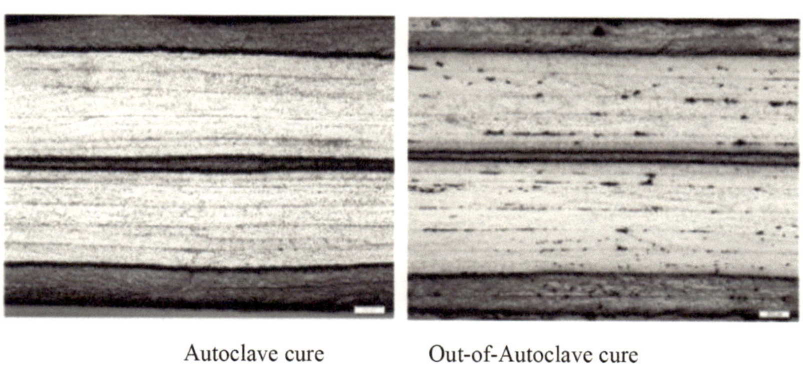

Figure 12. Comparison of autoclave and out-of-autoclave processing.

Autoclave process is the major cost in composite manufacturing. The investment of the large size autoclave oven is formidable for any company. Boeing employed OOA resin matrix, and the tank exhibited approximately 3% porosity (**Figures 11** and **12**).

For further evaluation of OOA processing, two-hybrid laminates fiber-placed panels produced at Boeing and LH2 tested at MSFC. The laminates were made up of 12 plies of 0.137 mm and 5 plies of 0.064 mm material. The OOA laminates exhibited approximately 4% porosity.

Figure 13. Fluted-core skirt.

Figure 14. Four different composite tanks of CHATT project.

Testing with autoclave coupons and the same materials did not show measurable permeability. Evaluation of the OOA laminates revealed that microcrack forms in the thin plies, primarily due to the porosity in the laminate. To deduce porosity, and eliminate permeability for OOA, it is necessary to increase the number of thin plies and to reduce porosity by improving the OOA materials architecture and fiber placement processes.

(4) Robotic AFP

Instead of the gantry-based automated fiber placement (AFP) equipment classically used to build large cylindrical parts, Boeing opted to use robotic fiber placement (AFP). Robotic AFP system enabled improved capabilities–better reach in the dome areas and lower shaft clearance.

(5) Fluted-core composite skirt

To prevent the delaminations occurred in X-33 composite cryotanks, Boeing opted an innovative fluted-core skirt, comprising large trapezoidal members-technically, laminate-angled web members with structural-radius fillers between facesheets [24]. The fluted-core skirt was designed to take the compressive load during launch and to vent permeated hydrogen. The skirt structure was co-cured to eliminate potential de-bonding issues at the working temperature of 253°C (**Figure 13**).

2.5. CHATT project

The project Cryogenic Hypersonic Advanced Tank Technologies (CHATT) is a part of the European Commission's Seventh Framework Programme and run on behalf of the Commission by DLR-SART in a multinational collaboration. The main objective was to investigate carbon fiber reinforced composite material for cryogenic fuel tank applications (**Figure 14**).

Four different subscale CFRP-tanks have been designed, manufactured, and tested. The CHATT project contributed to significant progress in the design of composite tanks for cryogenic propellant applications in Europe. The European Technology Readiness Level (TRL) of such cryotank is in the range between three and four, while the TRL in the US is considerably more advanced [25].

3. LOX cryotank

The weight reduction provided by composite Lox tank is too great to disregard. The most difficult issue for composite Lox tanks is material compatibility. After solving the problems of the hydrogen permeation and cryogenic properties of LH2 tank, MDA and NASA investigated the compatibility of composite materials and liquid oxygen (LOx). The test results proved that composites could be used to fabricate Lox tanks for launch vehicles, and the results sufficiently convincing to plan on building and flying composite Lox tanks in X-33 and X-34 vehicles [26].

NASA and Lockheed Martin jointly developed the first composite cryogen tank for liquid oxygen storage. Lockheed Martin designed and manufactured a sub-scale LOX cryotank, while NASA tested it at its MFSC facility.

The composite tank is approximately 2.7 m in length and 1.2 m in diameter, and weighs less than 225 kg, the weight saving is 18% compared to metal tank of similar construction. The

Figure 15. Composite lox tank of LM.

Figure 16. The 12-m diameter composite lox tank of space X.

tank passed the initial testing of cyclic liquid oxygen loading. The tank withstood the thermal and pressure conditions similar to that on a space launch vehicle. The tank did not permeate nor crack after 52 cycles of fill-drain liquid oxygen test, and passed the demonstrated test, as shown in **Figure 15** [27].

This tank marks a real advance in space technology. No approved standards for composite pressure vessels exist; there has not been enough information on them to write standards. So the technical data getting from this effort is very valuable [24].

Space X has designed, developed, and fabricated a composite LOx tank, which is a key component of interplanetary transport system (ITS). The tank is 12 m in diameter, and passed 2/3 exploration pressure test successfully in Nov 2016, see **Figure 16**. The carbon fiber used in the tank was provided by Toray. The 12-m diameter tank was the largest vessel ever produced [27].

http://dx.doi.org/10.5772/intechopen.73127

4. Materials

Fiber-reinforced composites can be optimized and tailored with the right amount of fibers based on the directions and magnitudes of the stress state. Composite cryotanks can be the most efficient isotensoid structure. Yet the challenges remain with carbon fibers particularly for manufacturing large-scale cryotank.

IM7 which is manufactured by Hexcel, was used widely in cryogenic tank in NASA. IM600 which is made by Toho, was used in cryotank in Japan [28]. T800H which is made in Toray, its property parameters are equivalent to IM7 and IM600 (**Table 2**). All of them are intermediate modulus fibers, high tensile strength, PAN-based fiber, their properties are basically equivalent.

The carbon fiber with intermediate modulus, high tensile strength was used widely in composite cryotank manufacturing. The higher the modulus of carbon fiber, the higher the crystallinity of the fiber, this would lead to reduced surface active functional groups and decreased the interlaminar shear strength of composites. Low modulus carbon fiber is not good for

	Fiber type	**Tensile strength/ MPa**	**Tensile modulus/ GPa**	**Strain/%**	**Density/g/cm³**	**Filament diameter/μm**
Toho	IM600	5790	285	2.0	1.80	5.0
Hercules	IM7	5300	275	1.8	1.77	5.2
Toray	T800H	5490	294	1.9	1.81	5.0

Table 2. Property parameters of carbon fibers.

	IM7/5320-1 [31]	**IM7/977-2 [32–34]**	**IM7/8552 [34]**
0°Tensile strength/MPa	2703	2690	2650
0°Tensile modulus/GPa	156	165	168
Poisson's ratio	0.34	—	—
90°Tensile strength/MPa	81	75	—
90°Tensile modulus/GPa	9.7	7.6	—
0°Compressive strength/MPa	1737	1580	1690
0°Compressive modulus/GPa	143	152	150
Short beam shear strength/MPa	119	112	128
CAI/MPa	176	262	234
Open-hole tensile strength/MPa	498	448	—
Open-hole compressive strength/MPa	386	310	—

Table 3. Properties parameters of IM7/5320-1and IM7/977-2 laminates.

the mechanical properties of composites. Therefore, intermediate modulus carbon fiber was employed for cryotank application [29, 30].

Toughened resins were developed primarily for aircraft applications to improve the compression-after-impact strength (CAI) of composite structures. They are also preferred for liquid hydrogen tanks because of greater impermeability after thermo mechanical cycling. For example, 8552 epoxy resin, which was made by Hercules, 977-2 and 5320-1resins, which were made in ICI, were used widely in LH2 cryotank fabrication.

Toughened resins appeared to be generally more resistant to ignition; therefore, toughened resins were used in LO2 tank fabrication. LOx compatibility is another issue that should be considered for Lox tank.

CF/977-2 prepreg was used widely in cryogenic applications because of excellent toughness and processing, for example, X-33 LH2 tank and 10 m composite demonstrator tank made in CCTD phase I [31, 32]. CYCOM 977-2 is a 177°C curing toughed epoxy resin. It is formulated for autoclave or press molding. Unidirectional tape and woven fabric impregnated with CYCOM 977-2 resin will retain tack for at least 10 days and has a long mechanical out life suitable for fabrication of large structures.

Table 3 showed the properties of IM7/ 5320-1, IM7/977-2, and IM7/8552 laminates. Tensile property, compressive property, and shear property of them were basically equivalent, the CAI of IM7/ 5320-1 were the lowest. The advantage of 5320-1 is that it requires no autoclave curing.

5. Conclusions

Composites are seen as one of the key components in the drive by NASA and the aerospace industry to decrease the weight of future launch vehicles as a means of reducing the cost of launching payloads.

NASA and Boeing have made significant progress in the US within the Composite Cryotank Technology Demonstration (CCTD) Project in which a large-scale composite liquid-hydrogen cryogenic tank has been designed, built, and tested under relevant flight loads.

For future space missions, there is still work to be done in producing out-of-autoclave tanks that meet the stringent permeability requirements.

Author details

Hongfei Zheng[1]*, Xuesen Zeng[2], Jianbao Zhang[1] and Hongjie Sun[1]

*Address all correspondence to: hongfei_0213@163.com

1 Aerospace Research Institute of Materials and Processing Technology, Beijing, China

2 University of Southern Queensland, Centre for Future Materials, Toowoomba, Queensland, Australia

References

[1] John V. Composite Cryotank Project Structures for Launch Vehicles, Composites Australia Conference; 2013

[2] Robinson MJ, Stoltzfus JM, Owens TN. Composite Material Compatibility with Liquid Oxygen. AIAA-97-1107

[3] Robinson MJ. Composite Cryogenic Propellant Tank Development, AIAA-94- 1375-CP

[4] Robinson MJ, Stoltzfus JM, Owens TN. Composite Material Compatibility with Liquid and Gaseous Oxygen, AIAA-2001-1215

[5] Robinson MJ, Eichinger JD, Johnson SE. Hydrogen Permeability Requirements and Testing for Reusable Launch Vehicle Tanks, AIAA 2002-1418

[6] Wilkerson C. Acoustic Emission Monitoring of the DC-XA Composite Liquid Hydrogen Tank During Structual Testing, NASA Technical Memorandum 108520

[7] DumbacherD, Results of the DC-XA Program, AIAA-96-4317

[8] Robinson MJ, Johnson SE, Eichinger JD, Hand ML, and Sorensen ET. Trade Study Results for a Second-Generation Reusable Launch Vehicle Composite Hydrogen Tank, AIAA 2004-1932

[9] Wright RJ, Roule GM. LH2 Tank Composite Coverplate Development and Flight Qualification for the X-33

[10] Glaessgen EH, Reeder JR, Sleight DW, Wang JT, Raju IS, Harris CE. Debonding failure of sandwich-composite cryogenic fuel tank with internal Core pressure. Journal of Spacecraft and Rockets. 2005;**42**(4)

[11] NaftelC. X-33, Stepping Stone to Low Cost Access to Space

[12] HodgeAJ. Evaluation of Microcracking in Two Carbon-Fiber/Epoxy-Matrix Composite Cryogenic Tanks. NASA/TM-2001-211194

[13] Mindy N. X-33 LH2 Tank Failure Investigation Findings, Manufacturing Problem Prevention Workshop; 2002

[14] AbumeriGH, KosareoDN, RocheJM. Cryogenic Composite Tank Design for Next Generation Launch Technology; AIAA

[15] WarwickG. Leak-proof Composite Tank Could Hoid Fuel or Astronauts, Flight International, Sep 14-Sep 20, 2004;166,4951; ProQuest

[16] Composite fuel tank passes NASA tests, February 2004 REINFORCEDplastics

[17] Composite Cryotank Technologies & Demonstration, Game Changing Development Program Office, NASA

[18] Composite Cryotank Technologies & Demonstration 2011-2014, NASA

[19] John V. NASA Composite Cryotank Technology Project Game Changing Program, 01/12/2015

[20] TF Johnson, DW Sleight, RA Martin, Structures and Design Phase I Summary for the NASA Composite Cryotank Technology Demonstration Project, AIAA 2013-1825

[21] Knapschaefer J. NASA/Boeing Composite Launch Vehicle Fuel Tank Scores Firsts, Composites World, 10/19/2017

[22] Jackson JR, John V, John F. Composite Cryotank Technologies and Development 2.4 and 5.5m Out of Autoclave Tank Test Results

[23] Rahmani N, Willard B, Lease K, Legesse ET, Soltani SA, Keshavanarayana S. The effect of post cure temperature on fiber/matrix adhesion of T650/Cycom 5320-1 using the micro-droplet technique. Polymer Testing. 2015;**46**:14-20

[24] Vickers JH, Tate LC, Gaddis SW, Neal RE. Composites Materials and Manufacturing Technologies for Space Applications, NASA/CP-2016-218217

[25] Sippel M, Kopp A. Progress on Advanced Cryo-Tanks Structural Design Achieved in CHATT-Project

[26] Grayson GD, Cook LM. Performance Characteristics of the DC-XA Liquid Oxygen Propellant-Acquisition System, AIAA 96-3081

[27] Musk E. Making Humans a Multiplanetary Species, 67th IAC, Guadalajala, Mexico,9/2016

[28] Aoki T, Ishikawa T, Morino Y. Overview of Basic Research Activities on Cryogenic Composite Propeppant Tanks in Japan, AIAA-2001-1878

[29] Yu B, Liu ZD, Jin QC, Cheng B, Chen W, Shi XQ, Li XL. The review of world-wide space system composite pressure vessel and the development trend analysis. Pressure Vessel. 2012;**29**(3)

[30] Cheng H, Debo L, Wu HG, Zhilong C. Application prospects of composite propellant tanks in domestic launch vehicles. Journal of Shenyang Aerospace University. 2016;**33**(2)

[31] CYCOM® 5320-1 EPOXY RESIN SYSTEM TECHNICAL DATA SHEET

[32] CYCOM® 977-2 EPOXY RESIN SYSTEM TECHNICAL DATA SHEET

[33] Pavlick MM, Oliver MS, Johnson WS. Determination of Interlaminar Toughness of IM7/977-2 Composites at Temperature Extremes and Different Thicknesses

[34] Zhao Q. Advanced Composite Handbook. Beijing: China Machine Press; 2003. pp. 408-410

Chapter 7

Clathrate Hydrates

Juwon Lee and John W. Kenney III

Additional information is available at the end of the chapter

http://dx.doi.org/10.5772/intechopen.72956

Abstract

The clathrate hydrates represent a distinctive, unusual, scientifically significant, and practically important class of solid state materials. Since their discovery in the early nineteenth century, their widespread distribution in oceans and permafrost regions and their ability to trap atoms and small molecules—particularly methane and other small hydrocarbons—has led to the realization that they are simultaneously a tremendous energy source and, in the face of global warming, a potential greenhouse gas release disaster of unprecedented magnitude just waiting to happen. In the twentieth century, it was realized that solid methane clathrate hydrate could plug natural gas pipelines and disrupt oil drilling processes. On the environmental positive side, clathrate hydrates can store hydrogen and sequester carbon dioxide. A brief historical review of the formation, structure, and uses of clathrate hydrates forms the backdrop for a discussion of modern scientific investigations of these solids employing spectroscopy, structure determination methods, isotopic studies, computational-theoretical modeling, and interrogations of guest-host interactions via special guests. For example, the use of colored halogens in clathrate hydrate hosts enables UV-visible spectroscopic methods to be employed to study clathrate hydrate structure.

Keywords: clathrate, clathrate hydrate, guest, host, lattice clathrate, methane clathrate hydrate, halogen clathrate hydrate, alkane clathrate, hydrogen bonding network, greenhouse gas, enclathration, sequestration, solidification

1. Introduction

Pure water solidifies (i.e., freezes) at 0°C at the normal atmospheric pressure of 1 bar. Pure chlorine gas solidifies at −101°C at ambient 1 bar pressure. However, as Humphery Davy [1] famously noted in 1811 "the solution of oxymuriatic gas (chlorine) in water freezes more readily than pure water." It turns out that chlorine-water solutions solidify at temperatures well above 0°C. This phenomenon is by no means restricted to chlorine-water solutions. Aqueous solutions into which many types of small non-polar or modestly polar gas atoms or molecules are dissolved (e.g., xenon, bromine, iodine, oxygen, nitrogen, carbon monoxide, carbon dioxide, methane, ethane) exhibit this enhanced solidification temperature effect also. Application of modest pressures further increases the solidification temperatures of these solidified gas-water solutions. For example, at 50 bar, a methane-water solution solidifies around 8°C. It turns out that the gas molecules or atoms become guests encased in a three-dimensional lattice network of water molecules called a clathrate hydrate host. Many clathrate hydrate host cage structures, sizes, and geometries are possible. If the guest molecule is methane and the host system is water, the solidified guest-host system is called methane clathrate hydrate. The interaction between the host cage water molecules in the clathrate hydrate and the guest gas atoms or molecules is of the non-covalent van der Waals type. The distinctive and fascinating solidification behavior of gas-clathrate hydrate systems is the subject of this chapter.

The scientific history of clathrates dates back to the year 1811 when Sir Humphrey Davy first studied and identified clathrates comprised of solidified mixtures of gases, particularly methane, and water [1]. A distinctive property of these natural gas clathrate hydrates is their capacity to remain in the solid state well above the 273.15 K melting point of pure water. A quantitative study of the composition of gas clathrate hydrates was later taken up by Michael Faraday [2]. After reading Schroeder's 1927 review of clathrate hydrates [3], Hammerschmidt [4] in the mid-1930s came the realization that solidified methane clathrate hydrates were responsible for the blocking of natural gas transmission pipelines, a major problem in the petroleum industry. From the 1930s to the present, the petroleum industry has expended an enormous amount of effort in developing techniques to detect and remove/dislodge solid gas clathrate hydrates or prevent their formation. It is now known that massive amounts of methane are sequestered in clathrate hydrates in the deep ocean and permafrost regions of planet Earth. This enclathrated methane has the potential to be extracted and used as a fuel. On the negative side, global warming will increase ocean temperatures and destroy permafrost regions, resulting in a potentially catastrophic release of methane – a major greenhouse gas – into the atmosphere. It is speculated that clathrate hydrates are widespread in the universe, and possibly exist in comets, asteroids, Mars, Pluto, and the moons of Jupiter and Saturn in our Solar System [5].

Clathrates are defined by the IUPAC to be inclusion compounds in which the guest atoms or molecules are constrained or trapped in cages formed by the host molecule or by a lattice of host molecules. Host molecules and guest atoms or molecules interact via van der Waals or hydrogen bonding forces. The word clathrate comes from the Latin word clatratus, which means a lattice or with bars. The traditional use of the word clathrate in science and engineering is in reference to materials comprised of hosts that are polymers trapping guests that are atoms or molecules. However current use of this word now extends to many other molecular host systems such as calixarenes [6], cyclodextrins [7], and even to inorganic polymeric systems such as zeolites [8] and chibaites [9]. Hofmann compounds [10] and metal organic frameworks (MOFs) [11] can also serve as clathrate hosts for small aromatic guest molecules such as carbon tetrachloride, benzene, toluene, and xylene. These types of clathrate host systems have the potential to serve as a new class of stationary phases capable of achieving heretofore unattainable chromatographic separations. Clathrates may also be separated into two general types depending upon how the host molecules encapsulate the guest atoms or molecules. In molecular clathrates, each guest atom or molecule is trapped in a site located in a single host molecule. Such clathrates are stable in both the liquid solution and solid states. However, in lattice clathrates, the trapping sites for guests are formed from multiple host molecules arranged in a lattice or array. Thus, lattice clathrates only exist in the solid state since dissolving or melting destroys these intermolecular host lattice trapping sites. Perhaps the most famous and well known example of a lattice clathrate is methane clathrate hydrate. Each methane guest molecule is trapped in a cage formed from a three-dimensional network of hydrogen-bonded host water molecules. It turns out that there are many other examples of clathrate host molecules that form hydrogen bonded networks capable of trapping guest atoms and molecules. The compounds hydroquinone, urea, thiourea, and Dianin's compound, (4-p-hydroxyphenyl-2,2,4-trimethyl chroman) are notable examples [12, 13]. The clathrate hydrate host lattice can form hydrogen bonded water molecule cages of various sizes and geometries capable of trapping a wide variety of guest atoms and molecules. Examples of clathrate hydrate guest moieties include a number of small alkanes, hydrogen, the noble gases, the halogens, nitrogen, oxygen, and carbon dioxide, carbon monoxide, and ozone.

2. Clathrate hydrate applications and uses

2.1. Petroleum industry

Clathrate hydrate plugs in gas pipelines have been very problematic for energy industries, causing significant economic loss and environmental risks. In 2010, the Macondo Project Deep Horizon well blew up and created an oil spill of 4.9 million barrels in Gulf of Mexico, which simultaneously released large amounts of hydrocarbons [14]. The hydrocarbons that were released ended up forming methane clathrate hydrates with surrounding water vapors. Unfortunately, the accumulation of methane clathrate hydrates blocked the well caps and prevented a 100-ton containment structure from being sealed after the blowout (**Figure 1**).

2.2. Environmental issues

A long-term concern for natural clathrate hydrates is the slow release of methane gas into the atmosphere [15]. For instance, when oil companies drill hot oil through hydrate-bearing

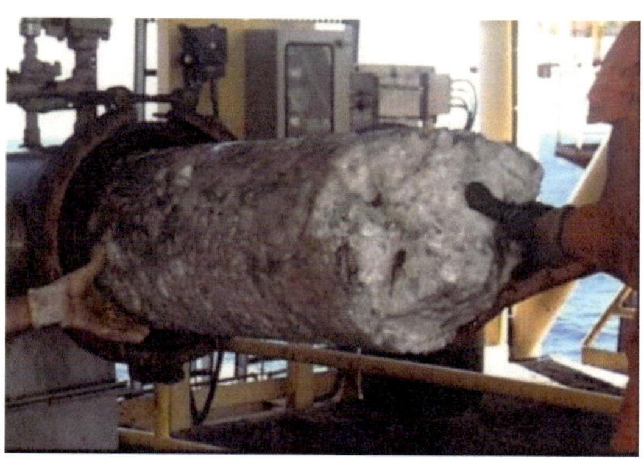

Figure 1. Recovered gas hydrate plug from a subsea pipeline off the coast of Brazil (courtesy of Petrobras).

sediments, the high temperature of the oil encourages melting and thus dissociation of clathrate hydrates. Then, large amounts of methane gas are released, surrounding sediments are weakened, and pockets of highly pressured gas are generated. The effect of global warming on the stability of methane clathrate hydrate is another concern. The rising temperature results in a steady release of methane, a key greenhouse gas, from methane clathrate hydrates in the Arctic where there are large deposits of methane clathrate hydrates. The dramatic dissociation of methane hydrates, also known as the clathrate gun hypothesis, exacerbates global warming and poses a significant threat to our climate [16]. Recently, large craters and blowouts from the Barents Sea floor that are produced from decomposing methane hydrates in the ocean floor have been discovered [17, 18].

2.3. Clathrate hydrate as energy source

Clathrate hydrates have been also studied as a potential energy source and fuel storage/transportation technology. Moreover, they are readily available; there exist extensive clathrate hydrate deposits along continental margins and in the Arctic. In particular, methane clathrate hydrates have been found in deep ocean layers, sediments below the Arctic, and permafrost areas where there is a combination of high pressure and low temperature. A reliable estimate done by DOE and USGS researchers suggests that there are about 3×10^{15} cubic meters of methane or ~1800 gigatons of carbon in hydrates that can be utilized as a potential energy source. In the US alone, the mean estimate of the amount of gas sequestered in clathrate hydrates is around 9×10^{21} m^3 [19]. Another USGS survey indicates that the potential energy stored in methane clathrate hydrates exceeds the combined global fossil fuel reserves! Thus, natural gas stored in the form of clathrate hydrates may be a significant alternative energy source in the future (**Figure 2**).

Extracting combustible gases, especially methane, already stored in natural clathrate hydrates is a prospect, but using the clathrate hydrates to hold gas molecules that can serve as fuels is also a subject of study. Storing gases in a solid state clathrate hydrate "container" may be more cost-effective than storage in the liquid or gaseous state in many cases. For instance, building a LNG plant is much more costly than storing natural gas in clathrate hydrates. Clathrate hydrates also offer an easier method of transportation compared to pipelines in many instances. Clathrate hydrates are also able to sequester hydrogen, the ultimate, carbon-free, environmentally friendly fuel. In the past decades, researchers have tested different types of promoters to increase hydrate formation rate and decrease induction time in order to enhance clathrate hydrates as a storage method.

Large scale trapping of carbon dioxide molecules in clathrate hydrates is a potentially viable way to reduce atmospheric CO_2 levels and thus inhibit the acidification of ocean waters [20]. Other applications of clathrate hydrates include the separation of gases such as CO_2 from flue gases and desalination. Clathrate hydrates can be used in flue gases to separate CO_2 by encouraging the formation of CO_2 clathrate hydrate in a flue gas mixture. Seo et al. [21] found that forming clathrate hydrates is be a feasible method of recovering CO_2 from flue gases in silica gel pores. Another useful application is hydrate-based desalination. The process begins when clathrate hydrate forming agent is injected into seawater that has a surrounding temperature lower than clathrate hydrate forming temperature. The condition promotes

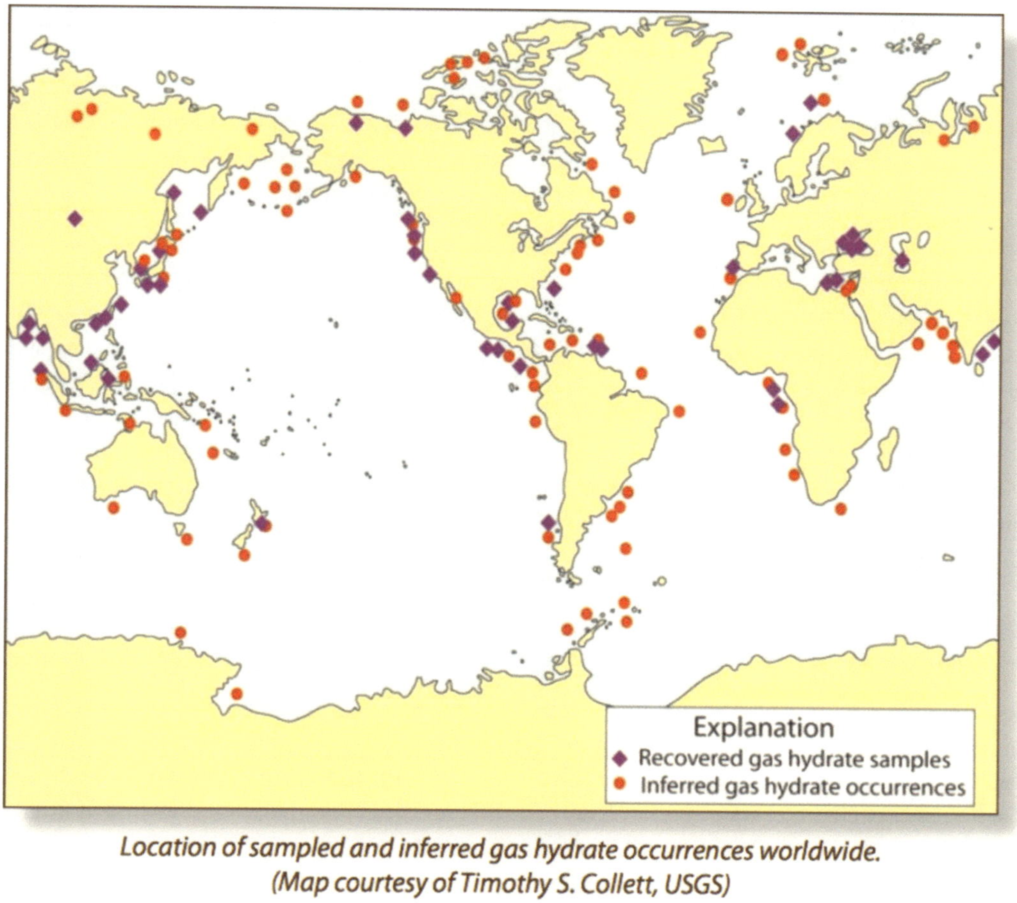

Figure 2. World map of recovered and inferred gas hydrate locations (map courtesy of Timothy S. Collett, USGS).

condensation of water molecules around the hydrate formers, and a slurry of clathrate ice and brine form. Once they form, brine is separate from the slurry of clathrate ice. The clathrate ice is then melted via heat exchange with warmer surface water of the ocean. During the last decade, the United States, Canada, Japan, India, and South Korea have conducted extensive research on clathrate hydrates. While much is now known, much still remains to be discovered. To advance understanding concerning the physical properties and practical uses of clathrate hydrates, further research is necessary.

3. Clathrate hydrates in space

The possible presence of clathrate hydrates in Mars, Pluto, Saturn's and Jupiter's moons, Kuiper Belt objects, comets, and asteroids is a hotly debated topic. In the cosmos, the giant planets may provide a cold, gaseous environment for clathrate hydrates to form as grains of water ice and gaseous guest molecules collide. Primordial sources of volatiles in space can be found in clathrate hydrates that formed in the presolar nebula. They have remained stable in low pressure-low temperature condition. Based on current information, it is speculated that natural gas hydrates may exist on Mars, Saturn, Uranus and Neptune and their moons. During the past decades, ammonia and methane clathrate hydrates have been the focus of space research because they have been considered to be the main sources of nitrogen and carbon species.

Titan, one of Saturn's moons, has been extensively investigated for the presence of clathrate hydrates because of its likely conditions. Titan's atmosphere and surface contain atmospheric gases such as nitrogen, methane, ethane, carbon dioxide, propylene, water vapor that typically form clathrate hydrates. However, since the atmosphere has only a limited amount of water vapor presence despite an abundance of hydrate guests, the amount of available water regulates clathrate hydrate formation. Osegovic and Max [22] suggested that the overabundance of hydrate forming guest molecules in Titan may exhaust all water vapor supply on the surface. In the event that water is introduced to the surface such as in cryovolcanism, the newly introduced water vapor would most likely form clathrate hydrates with guest molecules in the atmosphere unless a more thermodynamically favorable water vapor sink consumes water molecules first. In Titan, compound clathrate hydrates may form in which two or more guest molecules are trapped in a single clathrate cage. These clathrates usually assume sII or sH type structures. They tend to have higher reaction rates and faster induction time compared to pure clathrate hydrates.

The most common type of clathrate hydrate conjectured to be stored below the surface of Titan is methane clathrate hydrate since methane gas is one of the most common types of gases in Titan, as discovered by the Cassini-Huygens space probe, and methane clathrate hydrates have a broad spectrum of stability. Loveday et al. [23] reported that methane clathrate hydrates undergo structural transition at about 1 and 2 GPa and are able to remain stable up to 10 GPa. They are very stable in high pressure settings up to 42 GPa as well. After accretion, Lunine and Stevenson have suggested that the condensed methane in Titan is stored beneath a pure rock layer [24]. The inner core (~20–35 kilobars) of Titan thermodynamically prefers low pressure phases such as methane clathrate and ammonia monohydrate rather than equilibrium ices. Considering methane clathrate hydrates' characteristics and Titan's environment, it is likely that methane has been stored in clathrate hydrates in Titan for a long time.

Recent studies have also discussed the presence of ammonia in the cosmos including two of Saturn's moons, Titan and Enceladus [25]. Ammonia has previously been considered to be water-ice antifreeze and methane hydrate inhibitor, contributing to decreases in dissociation temperature of methane clathrate hydrates. Structurally, ammonia is around the molecular size of methane. However, ammonia is hydrophobic, which is atypical of most clathrate guest compounds. Previous experiments have shown that low pressure ammonia dihydrate and high pressure phase II of ammonia monohydrate resemble semiclathrate structures. Hence, Ripmeester et al. [25] experimented with ammonia clathrate hydrates and suggested that low temperature methane gas helps stabilize ammonia hydrate. Conversely, ammonia has been found to help activate ice nearby to initiate ammonia clathrate hydrate formation at temperatures where methane by itself does not normally form clathrate hydrates. They concluded that ammonia's inhibition effect on methane clathrate hydrates is not due to ammonia's instability as a guest molecule but instead due to their stabilizing effect of guest molecules at the aqueous phase [25].

http://dx.doi.org/10.5772/intechopen.72956

4. Clathrate hydrate crystal structure types and physical properties

As mentioned earlier, clathrate hydrates are nonstoichiometric, solid lattice inclusion compounds with a crystalline water cage that enclathrates small gas molecules. Generally there are vacant cages in the clathrate hydrates, but a sufficient amount of gas molecules are needed inside the cages to stabilize the structure. Typically, there is one molecule per cage. However in a high pressure environment [26] (>0.1 GPa), it is possible for a cage to contain more than one gas molecule. Thus, clathrates have similar physical properties as water including physical appearance, refractive index, and density. However, they also differ from water when considering properties like mechanical strength, heat capacity, thermal conductivity.

In the 1950s, von Stackelberg and Muller [27] were the first to determine a clathrate hydrate crystal structure using X-ray diffraction. They determined that CO_2 hydrates had the sI crystal structure. Since then, three most common crystal structure types of clathrate hydrates, have been identified by X-ray diffraction. The types are structure I (sI), structure II (sII), and structure H (sH). The sI structure is a body-centered cubic structure and has two cage types - a small pentagonal dodecahedral cage (5^{12}) and a large tetrakaidecahedral cage ($5^{12}6^{2}$). The structure sII is a diamond lattice within a cubic framework and has two cage types - small cage (5^{12}) and a large hexacaidecahedral cage ($5^{12}6^{4}$). The sII structure fits molecules in natural gases or oils that are larger than ethane but smaller than pentane. The sI and sII structures are of interest for gas companies because they form when natural gas molecules interact with

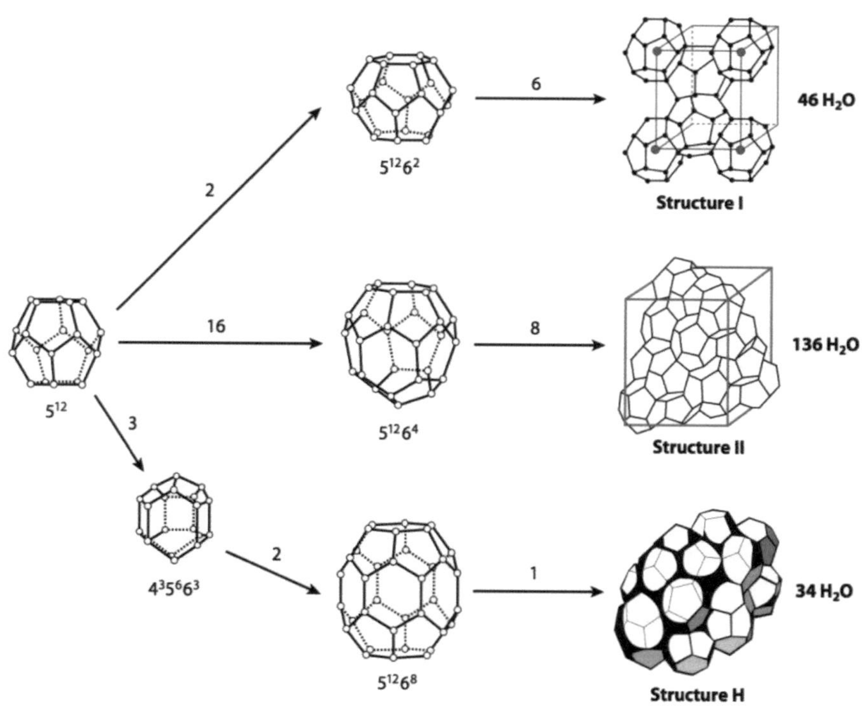

Figure 3. Common clathrate hydrate structures (sI, sII, sH) (courtesy of Koh et al.).

water near gas pipelines or deep ocean areas. The sH structure has a hexagonal framework and three cage types: a small 5^{12} cage, a mid-sized $4^{3}5^{6}6^{3}$ cage, a large icosahedral cage $5^{12}6^{8}$. Understanding of sH hydrate structure is particularly important for the oil industry because it is able to fit larger molecules contained in crude oils.

The cage size is determined by the size of the largest guest molecule. For example, methane can fit into both the small and large cages of sI whereas other large hydrocarbon molecules such as propane are too large to fit into the large cage of sI. Instead, they can fit into the larger cages of sII. In fact, clathrate hydrates formed near oil and gas pipelines are mostly of sII structure because they contain larger hydrocarbon molecules like propane and isobutane. Spectroscopy has been used to analyze guest occupancy. It has been found that molecules below 0.35 nm will not stabilize sI and above 0.75 nm will not stabilize sII [28]. When a large organic molecule is combined with a clathrate hydrate promoter, it is possible to form hydrates with atypical crystal structures. More new structures are being discovered as researchers produce clathrate hydrates with different types of gases in laboratory settings (**Figure 3**).

5. Memory phenomenon

Clathrate hydrate memory phenomenon suggests that clathrate hydrates nucleate faster from a dissociated hydrate solution than from a freshly made solution. It is an interesting phenomenon has been observed and recorded for over 100 years. However, some studies [29] have shown that memory phenomenon does not occur in all cases of clathrate hydrates. Thus, it has been suggested that heating the hydrate system at a high temperature can destroy the memory effect. However, even at high temperatures, researchers have observed the memory effect. Two possible explanations for the memory effect are either there is residual clathrate hydrate cage structure or clusters of high concentration gas left behind in the dissociated hydrate system. Buchanan et al. have observed that there are no significant signs of memory effect before and after the hydrate formation [30].

6. Models of thermodynamics, structure, and dynamics including solidification

Density functional theory (DFT) is the predominant method employed by researchers for modeling the structural, dynamical, thermodynamic, and spectroscopic properties of clathrate hydrates. Ramya and Venkatnathan have calculated interaction energies, free energies, and reactive energies of clathrate cages using DFT [31]. However, simulations based upon statistical mechanics, classical DFT, and Monte Carlo (MC) methods are also employed as shown in the work of Alavi and Ripmeester [32]. For example, Lasich and Ramjugernath used a simple lattice gas approximation with classical DFT coupled with Platteeuw theory to model clathrate hydrate phase equilibria for several hydrate guest molecules [33]. Standard DFT calculations were used by Chattaraj et al. to probe the structure, stability and reactivity

http://dx.doi.org/10.5772/intechopen.72956

of H_2 clathrates [34]. Using B3LYP HuzIII-su3 DFT, Siuda and Sadlej calculated NMR parameters (^{13}C and ^{1}H shielding constants and spin–spin coupling constants for 5^{12}, $5^{12}6^2$, and $5^{12}6^4$ cages hosting methane, ethane, and propane guests) [35]. Mondal and Chattaraj used ab initio molecular dynamics (AIMD) via atom-centered density matrix propagation (ADMP) methodologies to study time-dependent structural behavior of noble gas hydrates on the timescale of ~500 fs [36]. Sun et al. carried out DFT calculations using the M06-2X method to determine the relative stabilities of guest species in various clathrate hydrate host sites [37]. They predict that the guest molecules N_2 and NO are more stable in a $5^{12}6^4$ cage and that the SI NO hydrate is less stable than the SII NO hydrate. For the guest N_2, the Type II hydrate structure with single occupancy provides more stability than a type i structure with multiple occupancy. DFT calculations predict severe deformations when CS_2 is enclathrated in 5^{12} and $5^{12}6^2$ cages. The 5^{12}, $5^{12}6^2$, and $5^{12}6^4$, cages can enclathrate up to 2 N_2 molecules. Two H_2S molecules can be enclathrated in a $5^{12}6^4$ cage. DFT modeling and other types of modeling involving halogen guests in clathrate hydrate hosts are discussed in Section 10.

7. Clathrate hydrate structural studies

7.1. X-ray structural studies of clathrate hydrates

Much of the structural work on clathrate hydrates and their guests is based upon investigations of X-ray scattering from crystalline powders. Takeya et al. have determined structures of clathrate hydrates containing the guests CO_2, C_2H_6, C_3H_8, and methyl cyclohexane + CH_4 using X-ray methods with Rietveld analysis [38]. Shin et al. [39] have studied clathrate hydrate hosts with methanol guests via X-ray diffraction [39]. Davidson et al. carried out variable low temperature (18–250 K) powder X-ray investigations of THF clathrate hydrate from which its thermal expansion coefficient could be determined. Powder diffraction data of the CO_2--clathrate hydrate collected by Hartmann et al. obtained using synchrotron radiation as the X-ray photon source yielded structural information of extremely high precision via Rietveld refinements.

7.2. NMR, Raman, infrared spectroscopy

A diverse repertoire of spectroscopic techniques has been employed to investigate clathrate hydrate structure, bonding, composition, and dynamics. Chief among these techniques are X-ray crystallography, neutron diffraction, and infrared, Raman, and NMR spectroscopy. More exotic techniques include X-ray Raman and various time-resolved spectroscopies. For guest species that absorb near UV and visible light (e.g., the halogens), UV-visible absorption spectroscopy can be a useful tool by which to elucidate guest-host interactions. The interactions of clathrate hydrate hosts with xenon guests have been probed by ^{129}Xe NMR [48]. Similarly, ^{13}C NMR has been employed by Ripmeester and Ratcliffe [48] and Kida et al. [46] to investigate the dynamics of CO_2 and various alkane guests encapsulated in clathrate hydrate hosts. Chemical shift and linewidth variations can be correlated with the local guest-host interactions [42]. Infrared spectroscopy is a powerful tool for probing the compositions of extra-terrestrial objects. Evidence derived from laboratory infrared reflectance spectroscopy of the clathrate hydrate-CO_2 system at cryogenic temperatures by Oancea et al. [41] is inconsistent with corresponding reflectance spectra of Jupiter, Saturn, and the Martian poles suggesting that the observed spectral features do not come from clathrate hydrate-CO_2, although this does not rule out its existence beneath the surfaces of these bodies. Petuya et al. [43] have shown that Raman spectroscopy can be quite useful in determining the extent of partitioning of a CO guest in a clathrate hydrate host lattice containing several different host sites. The FTIR spectroscopy studies of Dartois et al. [40, 44] regarding the possibility of astrophysical remote detection of clathrate hydrates on Pluto and Triton lead to the observation that the methane (CH_4) associated with these bodies was not encapsulated in clathrate hydrates. Raman and resonance Raman investigations of halogen clathrate hydrates are most helpful in elucidating host-guest interactions in these highly colored systems.

8. Clathrate hydrates involving isotopic guests and hosts

Clathrate hydrate lattices formed out of ^{18}O and/or D are excellent model systems for probing the host lattice and guest vibrational behavior including the phonon density of states. A comparison of H_2O vs. D_2O as clathrate hydrate cage-forming molecules demonstrates that D_2O stabilizes the lattice better than H_2O. The crystalline D_2O clathrate hydrates are noticeably more stable thermodynamically than H_2O clathrate hydrates. The D_2O clathrate hydrates have assumed great importance in X-ray and neutron diffraction work in elucidating positions of the hydrogen atoms and thus the nature of hydrogen bonding in the clathrate hydrate structure. The use of isotopic substitutions (D vs. H and/or ^{18}O vs. ^{16}O) in varying proportions in clathrate hydrates provides the capability to adjust the clathrate host scattering intensity with respect to the guest scattering intensity in neutron inelastic scattering studies [45]. Isotopic substitution in alkane guests, particularly methane (e.g., $^{12}CH_4$, $^{12}CD_4$, $^{13}CH_4$, $^{13}CD_4$), provides unparalleled opportunities to investigate host-guest interactions through NMR, infrared, and Raman spectroscopies [46]. The NMR active ^{129}Xe guest has proven to be an exceedingly valuable as an NMR probe of clathrate hydrate properties [47, 48]. Through NMR, time-resolved probes of clathrate hydrates formation and decomposition can be studied; magnetic resonance imaging of clathrate hydrates is also possible [49]. Isotopes, even in an indirect way, can assume importance in studies involving clathrate hydrates. For example, observations of the isotopic content of deep ocean water indicate a direct correlation between ^{18}O concentration and the presence or one time existence of methane clathrate hydrates in deep ocean sediments [50].

9. Clathrate hydrate inhibitors and promoters

Inhibitors are chemicals that delay or prevent formation of clathrate hydrates under conditions where they would normally form. Promoters are chemicals that speed up the formation of and or stabilize the already formed clathrate hydrates under conditions where they might not normally be stable. There is tremendous incentive, particularly on the part of the petroleum and energy industries worldwide, to design better inhibitors and promoters that do not adversely affect the environment.

http://dx.doi.org/10.5772/intechopen.72956

9.1. Inhibitors

Clathrate hydrate formation is a crystallization process, and thus is affected by temperature, pressure, and other inhibiting or promoting factors. Clathrate hydrate growth inhibition has been a research focus since mid-1930s when, as previously noted, Hammerschmidt determined that man-made clathrate hydrates caused pipeline blockage. In order to inhibit clathrate hydrate formation, it is favorable to thermodynamically shift phase boundaries to high pressure and/or low temperature.

Modern thermodynamic programs based on a Gibbs energy extension of the van der Waals and Platteeuw method [51] calculation of hydrate stability temperature and pressure within a range. However, the low temperature and high pressure in the ocean where most pipelines are located create ideal hydrate-forming regions. To combat pipeline plugging, the oil industry has been injecting large amounts of hydrogen bonding fluids such as methanol or glycols that have antifreezing qualities. Introduction of hydrogen bonding fluids inhibits clathrate hydrate growth because they form hydrogen-bonds with available water, exhausting water from clathrate hydrate formation. Ultimately, it shifts the phase boundaries away from the hydrate forming zone. Methanol, in particular, has been widely used because it is easy to concentrate in water cages when injected as vapors. However, high concentrations (e.g., 40 vol%) of methanol pose a severe environmental threat. Large methanol contamination greater than 50 ppm in refinery feedstocks is severely fined [52]. Annually, in the US $220 million of methanol is being injected to inhibit hydrate growth (P.K. Notz). For instance, it requires about 4000 L = 4.0 m^3 of methanol to treat a gas pipeline that transports approximately 2430 m^3/d of gas. The methanol prevents clathrate hydrate formation but also carries the risk of contaminating the reflux water and increasing the potential for corrosion. The toxicity of methanol to humans and many other organisms is also a matter of grave concern.

Recently, hydrocarbon clathrate hydrate formation has been occurring in extreme environments that require a lot of methanol or other type of hydrogen bonding fluids, often too costly economically and environmentally, to successfully shift phase boundaries. Hence, much effort has been made to develop low-dosage hydrate inhibitors (LDHIs) in the past decade. The two type of LDHIs are kinetic hydrate inhibitors (KHIs) and anti-agglomerate chemicals (AAs). Their most significant benefit is that they pose little threat to the environment even in cases of leakages or mishandling. Thus, it is much safer to employ LDHIs in storage and transportation.

KHIs directly interfere in crystal lattice structures of clathrate hydrates, preventing crystal growth and nucleation, ultimately delaying or preventing hydrate formation. They are similar to enclathrated gas molecules and generally do not change the thermodynamics. They are supposed to be incorporated into growing clathrate hydrates and block further development. Some examples of polymeric kinetic hydrate inhibitors are poly(N-methylacrylamide), poly(N-ethylacrylamide), poly(2-ethyloxazoline) [53]. Recent studies have found that certain poly(vinyl caprolactam) (PVCap), a type of KHIs, changes the thermodynamics by increasing the dissociation temperature.

Anti-agglomerants (AAs) can be a solution to gas pipelines in very cold areas where thermodynamic inhibitor is impractical. AAs only work in the presence of an oil phase such as oil pipelines. They do not entirely prevent clathrate hydrates from forming but rather from agglomerating. AAs keep clathrate hydrate particles small and in a slurry so that they can be flushed out along with the fluids in the pipe. Many AAs have been made with halogen content. The high reactivity of halide compounds posed some threats with gas pipelines, and a lot of research has gone into developing low-halogen or no halogen AAs.

Considering long-term uses of inhibitors in gas pipelines, biodegradability has been one of the biggest concerns. As mentioned before, poly(vinylpyrrolidone) PVP and poly(vinyl caprolactam) PVCap are somewhat biodegradable. On the other hand, different types of esters such as pyroglutamic acid esters have high levels of biodegradability also have shown to function as gas hydrate inhibitors. Recently, researchers have been trying to create "bio-KHIs" [54] from natural living systems such as ocean pouts and mealworms that have proteins to help sustain frigid temperatures. The proteins may be applied to preventing pipeline plugging.

9.2. Promoters

In the recent decades, the use of clathrate hydrates as a storage and transportation method for gases such as methane, ethane, hydrogen, and carbon dioxide, has gained much attention. Unlike their liquid or compressed state counterparts, solid clathrate hydrates are much safer to handle and often cost-effective. Thus, many experiments have been conducted to examine promoters that increase induction time and growth rate and expand cage occupancy of clathrates. In essence, promoting additives work to drive clathrates away from the thermodynamic equilibrium or increasing reaction rates through efficient reactors. Some also work to increase stability of clathrate hydrates and increase clathrate occupancy. However, there have been concerns that promoter additives may take up space in the clathrate, limiting the occupancy for guest gas molecules.

Chemical additives that are generally considered as promoters are THF, surfactants, cyclopentane, acetone, propane. These types of additives do not alter the water cage structure. On the other hand, promoters like tetrabutylammonium salts and $[(n\text{-}C_4H_9)_4NBH_4]$ change the cage structure [53]. Recent studies have experimented with combining different promoters to discover synergistic effects that can drive clathrate hydrate growth as well as stability. Some methods [55] of increasing clathrate hydrate formation are vigorous mixing, using micron-sized ice particles or dry water, and using micron-sized ice particles, using hollow silica, metal packing media, porous materials to increases heat and mass transfer rate.

Tetrahydrofuran is a five-sided cyclic ether structure organic solvent that has been widely studied as a promoter. It has been generally considered a promoter due to its ability to form structure II hydrates with water and increase methane hydrate occupancy and stability [56]. THF additive has been observed to significantly reduce the hydrate formation pressure. The use of THF in methane hydrates has been suggested for upscaling the SNG technology as a more cost-effective and safer method to LNG. Furthermore, in methane clathrate hydrates (sII), THF increases the cage occupancy. Recently studies done by Veluswamy et al. [57] have reported that THF is an excellent kinetic and thermodynamic promoter for clathrate hydrate growth for CO_2 clathrate hydrates at moderate temperature (301.7 K) and pressure (1 MPa) of an unstirred tank reactor configuration. A potential application is using THF to improve natural gas storage in a solid state as clathrate hydrates. THF-promoted solid natural gas storage technique has potential to reduce significant cost in natural gas driven economies.

Surfactants are additives that reduce surface tension between two liquids or liquid and a solid, thereby increasing mass transfer. Surfactants do not change the thermodynamics of clathrate hydrates, but they have been found to alter clathrate hydrate kinetics. Karaaslan and Parlaktuna [58] have found that anionic surfactants increase hydrate formation rate at all concentration, cationic surfactants increase hydrate formation rate at low concentration, and nonionic surfactants have little distinguishable effects on hydrate formation rate. SDS is a commonly used surfactant that is known to increase hydrate stability. Arora et al. [59] conducted an interesting work developing a biosurfactant with *Pseudomonas aeruginosa* strain A11, which was characterized as rhamnolipids. 1000 ppm rhamnolipids solution increased methane hydrate formation and reduced methane hydrate's induction time. Biosurfactant research may lead to environmentally safe methods to use gas hydrates as an energy source.

Cyclopentane is a hydrophobic, cycloalkane compound. Lim et al. [60] conducted experiments comparing hydrate growth at different levels of cyclopentane, and observed that the system with 0.9 mL CP had a faster hydrate formation compared to the system with 0.45 mL CP. In the same experiment, it was noted that hydrates grow faster on metal surfaces due to better heat transfer. However, cyclopentane is not always effective as a clathrate hydrate promoter. Although cyclopentane is an efficient sII hydrate promoter at high temperatures (283 K), Herslund et al. [61] suggested that in its study of water-cyclopentane-carbon dioxide clathrate hydrate, cyclopentane's efficiency decreases at lower temperatures because cyclopentane clathrate hydrates are more likely to form than mixed carbon dioxide-cyclopentane hydrates at temperatures below 281 K and pressures above 0.4 MPa.

Tetra-n-butyl ammonium bromide (TBAB) is an additive that alters the clathrate hydrate cage and forms a semi-clathrate. In a semi-clathrate, the cage is broken so that the large TBAB molecule can fit inside. One advantage of semi-clathrates in storage purposes is that it allows for more guest molecules compared to promoters like THF, which do not alter the cage. Li et al. [62] examined a hydrate-based CO2 separation process from a fuel gas mixture and observed that TBAB range of 0.14–1.00 mol% increased the induction time and hydrate growth rate while also increasing CO_2 storage capacity of the clathrate.

An interesting phenomena similar to the effects of promoters that occurs in methane hydrates is self-preservation [55]. Self-preservation is when clathrates under ice layers form insulating ice shells that prevent dissociation. Thus, methane hydrate resists mass and heat transfer, allowing it to be kinetically stable in temperatures beyond the three equilibrium boundary.

9.3. Mixed promoters

Researchers have studied the presence of mixed promoters in clathrate hydrate formation to examine the synergistic effect on increasing hydrate formation rate and decreasing induction time. In a comparative study conducted by Kumar et al. [55], THF and SDS as mixed promoters for methane hydrates were tested. A concentration of 5.56 mol% THF was observed to induce the fastest rate of hydrate formation whereas 0.1 wt% of SDS was observed to increase the most gas storage capacity. Thus, a combination of THF and SDS has been suggested to optimize induction time and storage potential for methane hydrates. Partoon and Javanmardi [63] noted that a mixture of promoters, propanone and sodium lauryl sulfate (SDS), can work to increase induction time and initial formation rate of methane hydrates. However, the combination of the two promoters decreases self-preservation at low pressures and temperatures below the ice point and storage capacity of the hydrate. However, SDS by itself, expands the storage capacity compared to its theoretical value. The synergistic effect of mixed promoters needs to be further examined regards to cases in which a significant amount of pure promoter hydrates form at low temperatures.

9.4. Switching effects

Some additives have switching effects. For example, Mohammadi and Richon [64] found that while THF of 5–7 mol% has a promoting effect on clathrate hydrate formation, THF at high concentration (>10 mol %) in aqueous solution shifts phase boundaries to high pressure and thus plays an inhibiting role. Similarly, methanol and ethanol, typically inhibitors, have been found to act as a promoter for sII hydrates. Further research needs to be done on different types of additives that can either promote or inhibit hydrate formation depending on concentration and isotopic conditions.

10. Clathrate hydrate with halogen guests

Although halogen clathrate hydrates were among the first clathrate hydrates discovered, they have not attracted the level of scientific scrutiny that has been given to clathrate hydrates with hydrocarbon guests. Simply put, halogen clathrates do not plug natural gas pipelines. Recently, however, the value of studying this class of clathrates as model systems has been shown in a series of studies involving theoretical/computational modeling and experimental spectroscopy (**Figure 4**).

10.1. Theoretical and computational models of halogen clathrate hydrates

Over the past decade, theoretical and computational models of halogen clathrates hydrates of increasing sophistication have been utilized with considerable success to elucidate various aspects of halogen clathrate hydrate structure, thermodynamics, and spectroscopy.

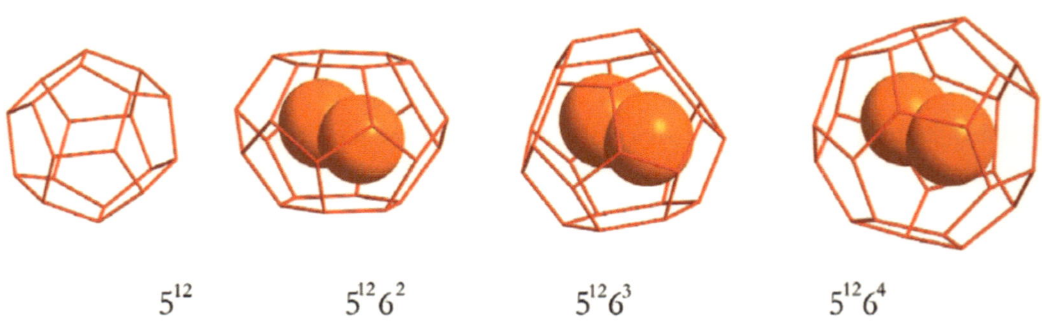

Figure 4. Common halogen clathrate hydrate structure types (courtesy of Janda et al.).

http://dx.doi.org/10.5772/intechopen.72956

Kerenskaya et al. [65] in 2007 used simple models and force field calculations to estimate the association energy between the clathrate hydrate host and the halogen guest. Schofield and Jordan [66] in 2007 carried out calculations modeling the ground and excited electronic states of Cl_2 trapped in 5^{12} and $5^{12}6^2$ clathrate hydrate lice atthost sites. A 700 cm^{-1} spectral blue shift was predicted for Cl_2 trapped in a distorted 5^{12} clathrate hydrate cage. Ramya and Venkatnathan [67] in 2013 employed DFT with a dispersion corrected (B97-D) functional to model the Raman spectra of H_2--a model system for the halogen X_2 guest molecules--encapsulated in pure and THF doped clathrate hydrates. Their computational results predicted a 700 cm^{-1} blue shift of the H_2 breathing mode in the THF doped clathrate hydrate guest vs. the pure clathrate hydrate. Molecular dynamics (MD) simulations, used by Nguyen and Molinero [68] in 2013 in conjunction with the coarse-grained model of water, were employed to model the thermodynamic stabilities of vacant and guest-filled sI, sII, TS, and HS-1 clathrate hydrate types. Bernal-Uruchurtu, Janda, and Hernandez-Lamoneda [69] developed a simple semi-empirical quantum mechanical model in 2015 incorporating, via molecular dynamics trajectories, the impact of movement of bromine molecules in clathrate cages on the UV-visible spectra of bromine clathrate hydrates. A simple electrostatic model of the clathrate hydrate cage predicts a blue shift in the UV-visible spectra, which agrees with experiment. Ochoa-Resendiz, Batista-Romero, and Hernandez-Lamoneda [70] in 2016 employed ab initio calculations and natural bond order analysis to investigate Cl_2 and Br_2 guests in various clathrate hydrate host sites. The picture that emerges is that of an interaction between the lone pair of the nearest oxygen atom in the clathrate hydrate host with the sigma LUMO of the halogen dimer X_2 but only when the halogen is out of its central equilibrium position in the cage, thus allowing a significant cage-halogen interaction. Breaking of O–H hydrogen bonds in the cage is not necessary in this interaction model. Dureckova et al. [71] in 2017 also used ab initio calculations and natural localized molecular orbital analysis in isolated cages and cages with one shell of coordinating atoms to elucidate halogen guest interactions with clathrate hydrate hosts.

10.2. Electronic and vibrational spectroscopy

Although halogen clathrate hydrates have been investigated extensively since their discovery in the early 1800s, there is a paucity of experimental spectroscopic data in the literature pertaining to the interaction between the halogen guest molecules and water host lattice in the clathrate. At this writing, the only halogen clathrate electronic and vibrational spectroscopic studies in the literature are the Raman investigations of Anthonisan [72], and the electronic (UV-visible absorption) and vibrational (Raman and resonance Raman) studies of Janda and co-workers [73]. Halogen clathrate hydrates absorb visible and near UV light but do not luminesce, even at cryogenic temperatures. The clathrate lattice efficiently dissipates the excited electronic state energy of the halogen chromophore through nonradiative channels. The UV-visible and Raman spectra measured Janda et al. [65, 74, 75] clearly show greater perturbations of the spectral properties of the enclathrated halogen guest when the water lattice host cage is smaller than when the lattice host cage is larger. For example, the UV-visible band maximum of the $X \rightarrow C$ transition of gas phase bromine is 24,270 cm^{-1} but this transition frequency increases to 24,630 cm^{-1} when the bromine is confined in a large $5^{12}6^4$ clathrate hydrate cage. Encapsulation of the bromine in a small $5^{12}6^2$ clathrate hydrate cage further increases the transition frequency to 25,150 cm^{-1}

(**Figure 5**). An almost identical frequency increase with decreasing guest cavity size is seen in the X → B electronic transition of bromine. This trend shows that tighter guest halogen confinement in a smaller host clathrate hydrate site--with consequently increased clathrate halogen interactions--also increases the B and C excited state energies relative to the energy of the X ground state. In effect, it is a particle in a box situation; a smaller box, in this case a smaller clathrate host lattice cage, gives rise to more widely spaced electronic energy levels and larger X → B and X → C transition frequencies (**Figure 6**). A molecular orbital explanation posits that an interaction between the halogen LUMO and the valence oxygen orbitals in the clathrate hydrate will result in a blue shift of the associated valence electronic bands of the water-encapsulated halogen. In contrast, the X → B transition frequency for gas phase chlorine is 30,300 cm^{-1} and for chlorine molecules trapped in small $5^{12}6^2$ clathrate cages this transition is also 30,000 cm^{-1}. This suggests that even the small $5^{12}6^2$ clathrate hydrate lattice minimally perturbs the valence orbitals of the encapsulated chlorine molecules, which are significantly smaller than bromine and iodine molecules. At the other extreme is iodine, the largest of the halogens studied by Janda et al. [75], whose gas phase X → B transition frequency of 18,870 cm^{-1} increases to 20,310 cm^{-1} when confined in a large $5^{12}6^4$ cage. The vibrational anharmonicity term w_ex_e in bromine clathrate hydrate decreases from 1.06 cm^{-1} in the gas phase, to 0.82 cm^{-1} in a $5^{12}6^2$ cage, and finally to 0.7 cm^{-1} in a larger $5^{12}6^4$ cage. This is consistent with the idea that the smaller and tighter the clathrate hydrate host trapping site becomes, the more harmonic the X ground state potential surface in the guest Br_2 becomes. The clathrate hydrate cage in effect "pushes on" the Br_2 potential surface via van der Waals interactions, rendering it more harmonic.

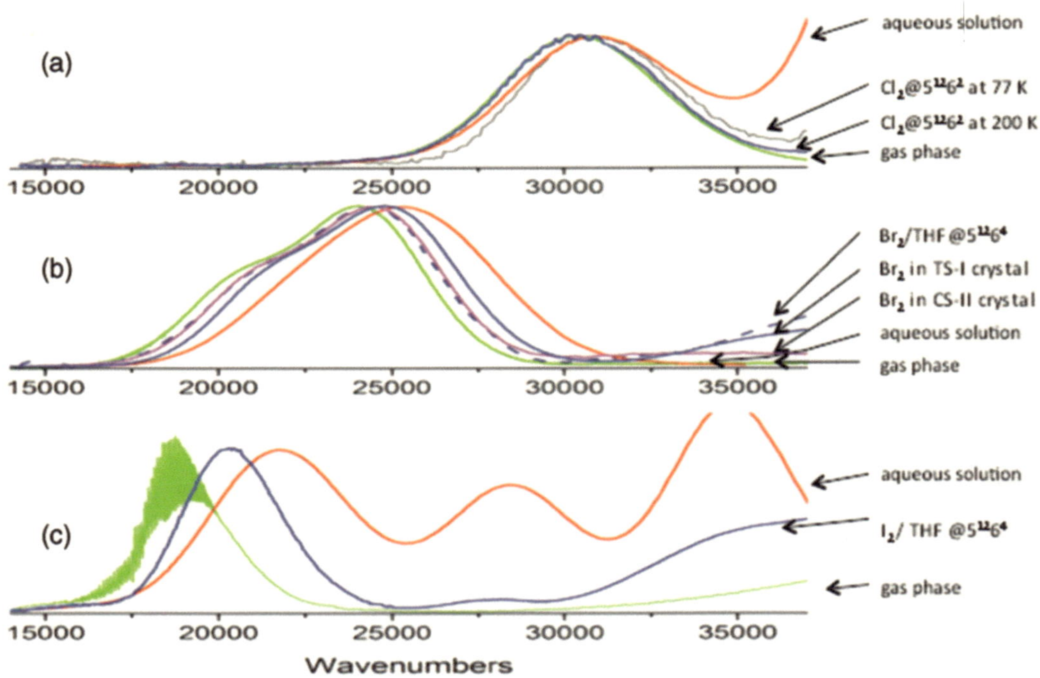

Figure 5. Spectra of (a) chlorine, (b) bromine, and (c) iodine in various environments. In each case, green curve is the gas phase spectrum and the red curve is for an aqueous solution. (Courtesy of Janda et al.).

http://dx.doi.org/10.5772/intechopen.72956

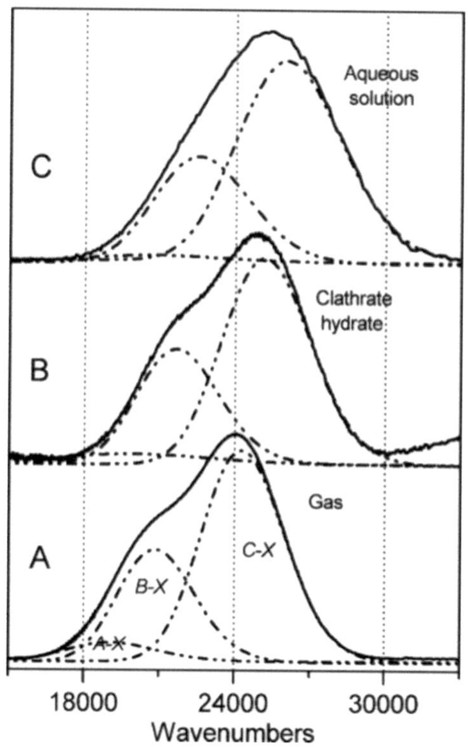

Figure 6. Br_2 absorption spectra in: (A) gas phase (293 K), (B) clathrate hydrate (250 K), and (C) liquid water (293 K). (Courtesy of Janda et al.).

11. Future research

Future research on encapsulated guests in clathrate hosts will undoubtedly focus to a certain extent on the design and execution of experiments that will yield definitive experimental confirmations of the existence of clathrate hydrates in other environments in the Solar System than Earth. Such experiments will undoubtedly include the landing of probe vehicles on the objects of interest in the Solar System. The probes will be capable of accessing, retrieving, and analyzing sub-surface samples where clathrate hydrates are thought to reside. Spectroscopic interrogations of extra-Solar planets will also include "clathrate hunting" in their task lists. On the practical side, the search for more effective and more environmentally friendly clathrate hydrate promoters and inhibitors will continue. Scientists and engineers will continue to study clathrate hydrates as possible storage/transport media for fuels such as methane and hydrogen. The possible use of clathrate hydrates as sequestering agents for carbon dioxide will continue to be investigated. Increasing scientific research attention will be given to the impact of global warming on methane clathrate hydrate degradation and subsequent release of methane into the oceans and atmosphere. In the realm of pure science, ever more sophisticated spectroscopic, structural, isotopic, thermodynamic, and theory/computational tools will be employed to provide a picture of unprecedented detail that encompasses the dynamics of clathrate formation, thermal degradation, and host-guest interactions. Expect also to see multi-spectroscopic/structural probes being used to investigate the same sample at essentially the same time under the same experimental conditions.

Acknowledgements

It is a pleasure to acknowledge Prof. Ken Janda for introducing us to the fascinating world of clathrate hydrate research.

Author details

Juwon Lee and John W. Kenney III*

*Address all correspondence to: john.kenney@cui.edu

Chemical Physics Laboratory, Concordia University, Irvine, CA, USA

References

[1] Sloan ED, Koh C. Clathrate Hydrates of Natural Gases. 3rd ed. Chemical Industries: CRC Press; 2007

[2] Faraday M. On fluid chlorine. Proceedings of the Royal Society of London. 1815;**2**:190-190

[3] Schroeder W. Die Geschichte der Gas Hydrate. Vortrage: Sammalung. Chem. Tech; 1927

[4] Hammerschmidt EG. Formation of gas hydrates in natural gas transmission lines. Industrial and Engineering Chemistry. 1934;**26**(8):851-855

[5] Landry JC, England AW. Far-infrared spectroscopy of CO_2 clathrate hydrate with Martian applications. Geophysical Research Letters. 1994;**21**:2829-2832

[6] Gutsche CD. Calixarenes. Vol. 50. Cambridge: Royal Society of Chemistry; 1989. pp. 293-294

[7] Menuel SÃ, Joly J-P, Courcot B, Elysée J, Ghermani N-E, Marsura A. Synthesis and inclusion ability of a Bis-Î²-Cyclodextrin pseudo-cryptand towards Busulfan anticancer agent. Tetrahedron. 2007;**63**(7):1706-1714

[8] Rollmann LD, Valyocsik EW, Shannon RD. Zeolite molecular sieves. Inorganic Syntheses. 1995;**22**:227-234

[9] Momma K, Ikeda T, Nishikubo K, Takahashi N, Honma C, Takada M, Furukawa Y, Nagase T, Kudoh Y. New silica clathrate minerals that are isostructural with natural gas hydrates. Nature Communications. 2011;**2**:196

[10] Culp JT, Madden C, Kauffman K, Shi F, Matranga C. Screening Hofmann compounds as CO_2 sorbents: Nontraditional synthetic route to over 40 different pore-functionalized and flexible pillared Cyanonickelates. Inorganic Chemistry. 2013;**52**(8):4205-4216

[11] Tranchemontagne DJ, Mendoza-Cortes JL, O'keeffe M, Yaghi OM. Secondary building units, nets and bonding in the chemistry of metal-organic frameworks. Chemical Society Reviews. 2009;**38**(5):1257-1283

[12] Nemkevich A, Spackman MA, Corry B. Mechanism of concerted hydrogen bond reorientation in clathrates of Dianinâs compound and hydroquinone. Journal of the American Chemical Society. 2011;**133**(46):18880-18888

[13] Bhatnagar VM. Clathrates of urea and thiourea. Journal of Structural Chemistry. 1968; **8**(3):513-529

[14] Deepwater Horizon – BP Gulf of Mexico Oil Spill. Available from: https://www.epa.gov/enforcement/deepwater-horizon-bp-gulf-mexico-oil-spill (Accessed Dec 29, 2017)

[15] Portnov A, Vadakkepuliyambatta S, Mienert JÃ, Hubbard A. Ice-sheet-driven methane storage and release in the Arctic. Nature Communications. 2016;**7**:10314

[16] Kennett JP, Cannariato KG, Hendy IL, Behl RJ. Methane Hydrates in Quaternary Climate Change: The Clathrate Gun Hypothesis: American Geophysical Union; 2003

[17] Solheim A, Elverhoi A. Gas-related sea floor craters in the Barents Sea. Geo-Marine Letters. 1993;**13**(4):235-243

[18] Andreassen K, Hubbard A, Winsborrow M, Patton H, Vadakkepuliyambatta S, Plaza-Faverola A, Gudlaugsson E, Serov P, Deryabin A, Mattingsdal R, et al. Massive blow-out craters formed by hydrate-controlled methane expulsion from the Arctic seafloor. Science. 2017;**356**(6341):948-953

[19] Collett TS, Lee MW. Reservoir characterization of marine and permafrost associated gas hydrate accumulations with Downhole well logs. Annals of the New York Academy of Sciences. 2006;**912**(1):51-64

[20] Nago A, Nieto A. Natural gas production from methane hydrate deposits using clathrate sequestration: State-of-the-art review and new technical approaches. Journal of Geological Research. 2011;**2011**:1-6

[21] Seo Y-T, Moudrakovski IL, Ripmeester JA, Lee J-W, Lee H. Efficient recovery of CO_2 from flue gas by clathrate hydrate formation in porous silica gels. Environmental Science & Technology. 2005;**39**(7):2315-2319

[22] Osegovic JP, Max MD. Compound clathrate hydrate on Titan's surface. Journal of Geophysical Research: Planets. 2005;**110**(E8):2156-2202

[23] Loveday JS, Nelmes RJ, Guthrie M, Belmonte SA, Allan DR, Klug DD, Tse JS, Handa YP. Stable methane hydrate above 2 GPa and the source of Titan's atmospheric methane. Nature. 2001;**410**(6829):661-663

[24] Lunine JI, Stevenson DJ. Clathrate and ammonia hydrates at high pressure: Application to the origin of methane on titan. Icarus. 1987;**70**(1):61-77

[25] Shin K, Kumar R, Udachin KA, Alavi S, Ripmeester JA. Ammonia clathrate hydrates as new solid phases for titan, enceladus, and other planetary systems. Proceedings of the National Academy of Sciences. 2012;**109**(37):14785-14790

[26] Hirai H, Tanaka T, Kawamura T, Yamamoto Y, Yagi T. Structural changes in gas hydrates and existence of a filled ice structure of methane hydrate above 40 GPa. Journal of Physics and Chemistry of Solids. 2004;**65**(8-9):1555-1559

[27] Stackelberg MV, Muller HR. On the structure of gas hydrates. The Journal of Chemical Physics. 1951;**19**(10):1319-1320

[28] Sloan ED. Gas hydrates: A review of physical/chemical properties. Energy & Fuels. 1998;**12**(2):191-196

[29] Wilson PW, Haymet ADJ. Hydrate formation and re-formation in nucleating THF/ water mixtures show no evidence to support a "memory" effect. Chemical Engineering Journal. 2010;**161**:146-150

[30] Buchanan P, Soper AK, Thompson H, Westacott RE, Creek JL, Hobson G, Koh CA. Search for memory effects in methane hydrate: Structure of water before hydrate formation and after hydrate decomposition. The Journal of Chemical Physics. 2005;**123**(16):164507

[31] Ramya KR, Venkatnathan A. Stability and reactivity of methane Clathrate hydrates: Insights from density functional theory. The Journal of Physical Chemistry. A. 2012; **116**(29):7742-7745

[32] Alavi S, Ripmeester JA. Simulations of hydrogen gas in clathrate hydrates. Molecular Simulation. 2017;**43**(10-11):808-820

[33] Lasich M, Ramjugernath D. Clathrate hydrates modelled with classical density functional theory coupled with a simple lattice gas and van der Waals-Platteeuw Theory. Philosophical Magazine. 2016;**96**(27):2853-2867

[34] Chattaraj PK, Bandaru S, Mondal S. Hydrogen storage in clathrate hydrates. The Journal of Physical Chemistry A. 2011;**115**(2):187-193

[35] Siuda PA, Sadlej J. Calculations of NMR properties for SI and SII clathrate hydrates of methane, ethane and propane. Journal of Molecular Modeling. 2014;**20**(12):2511

[36] Mondal S, Chattaraj PK. Noble gas encapsulation: Clathrate hydrates and their HF doped analogues. Physical Chemistry Chemical Physics. 2014;**16**(33):17943-17954

[37] Sun N, Li Z, Qiu N, Yu X, Zhang X, Li Y, Yang L, Luo K, Huang Q, Du S. Ab initio studies on the clathrate hydrates of some nitrogen- and sulfur-containing gases. The Journal of Physical Chemistry A. 2017;**121**(13):2620-2626

[38] Takeya S, Udachin KA, Moudrakovski IL, Susilo R, Ripmeester JA. Direct space methods for powder X-ray diffraction for Guestâ-host materials: Applications to cage occupancies

and guest distributions in clathrate hydrates. Journal of the American Chemical Society. 2010;**132**(2):524-531

[39] Shin K, Moudrakovski IL, Ratcliffe CI, Ripmeester JA. Managing hydrogen bonding in clathrate hydrates by crystal engineering. Angewandte Chemie International Edition. 2017;**56**(22):6171-6175

[40] Dartois E, Bouzit M, Schmitt B. Clathrate hydrates: Ftir spectroscopy for astrophysical remote detection. EAS Publications Series. 2012;**58**:219-224

[41] Oancea A, Grasset O, Menn EL, Bollengier O, Bezacier L, Mouélic SL, Tobie G. Laboratory infrared reflection spectrum of carbon dioxide clathrate hydrates for astrophysical remote sensing applications. Icarus. 2012;**221**(2):900-910

[42] Sengupta S, Guo J, Janda KC, Martin RW. Exploring dynamics and cage-guest interactions in clathrate hydrates using solid-state NMR. The Journal of Physical Chemistry B. 2015;**119**(50):15485-15492

[43] Petuya C, Damay FÃ, Talaga D, Desmedt A. Guest partitioning in carbon monoxide hydrate by Raman spectroscopy. The Journal of Physical Chemistry C. 2017;**121**(25): 13798-13802

[44] Dartois E, Bouzit M, Duret P, Schmitt B. Clathrate hydrate FTIR spectroscopy – infrared signatures, astrophysical significance. In: BIO Web of Conferences. Vol. 2. EDP Sciences; 2014. p. 03005

[45] Celli M, Colognesi D, Ulivi L, Zoppi M, Ramirez-Cuesta AJ. Phonon density of states in different clathrate hydrates measured by inelastic neutron scattering. Journal of Physics: Conference Series. 2012;**340**:012051

[46] Kida M, Sakagami H, Takahashi N, Nagao J. Chemical shift changes and line narrowing in ^{13}C NMR spectra of hydrocarbon clathrate hydrates. The Journal of Physical Chemistry A. 2013;**117**(20):4108-4114

[47] Ripmeester JA, Ratcliffe CI. Xenon-129 NMR studies of clathrate hydrates: New guests for structure II and structure H. The Journal of Physical Chemistry. 1990;**94**(25):8773-8776

[48] Ripmeester JA, Ratcliffe CI. The diverse nature of dodecahedral cages in clathrate hydrates as revealed by ^{129}Xe and ^{13}C NMR spectroscopy: CO_2 as a small-cage guest. Energy & Fuels. 1998;**12**(2):197-200

[49] Ripmeester JA, Ratcliffe CI. On the contributions of NMR spectroscopy to clathrate science. Journal of Structural Chemistry. 1999;**40**(5):654-662

[50] Hesse R, Harrison WE. Gas hydrates (clathrates) causing pore-water freshening and oxygen isotope fractionation in deep-water sedimentary sections of terrigenous continental margins. Earth and Planetary Science Letters. 1981;**55**(3):453-462

[51] Martin AÌ, Peters CJ. Hydrogen storage in SH clathrate hydrates: Thermodynamic model. The Journal of Physical Chemistry B. 2009;**113**(21):7558-7563

[52] Sloan ED. Fundamental principles and applications of natural gas hydrates. Nature. 2003; **426**(6964):353-363

[53] Eslamimanesh A, Mohammadi AH, Richon D, Naidoo P, Ramjugernath D. Application of gas hydrate formation in separation processes: A review of experimental studies. The Journal of Chemical Thermodynamics. 2012;**46**:62-71

[54] Sum AK, Koh CA, Sloan ED. Clathrate hydrates: From laboratory science to engineering practice. Industrial & Engineering Chemistry Research. 2009;**48**(16):7457-7465

[55] Kumar A, Kushwaha OS, Rangsunvigit P, Linga P, Kumar R. Effect of additives on formation and decomposition kinetics of methane clathrate hydrates: Application in energy storage and transportation. The Canadian Journal of Chemical Engineering. 2016;**94**(11):2160-2167

[56] Gough SR, Davidson DW. Composition of tetrahydrofuran hydrate and the effect of pressure on the decomposition. Canadian Journal of Chemistry. 1971;**49**(16):2691-2699

[57] Veluswamy HP, Premasinghe KP, Linga P. CO_2 hydrates effect of additives and operating conditions on the morphology and hydrate growth. Energy Procedia. 2017;**105**:5048-5054

[58] Karaaslan UÇ, Parlaktuna M. Surfactants as hydrate promoters? Energy & Fuels. 2000; **14**(5):1103-1107

[59] Arora A, Cameotra SS, Kumar R, Balomajumder C, Singh AK, Santhakumari B, Kumar P, Laik S. Biosurfactant as a promoter of methane hydrate formation: Thermodynamic and kinetic studies. Scientific Reports. 2016;**6**(1):1-13

[60] Lim Y-A, Babu P, Kumar R, Linga P. Morphology of carbon dioxide-hydrogen-cyclopentane hydrates with or without sodium dodecyl sulfate. Crystal Growth & Design. 2013;**13**(5):2047-2059

[61] Herslund PJÃ, Thomsen K, Abildskov J, Solms NV, Galfre AÃ, Brantuas P, Kwaterski M, Herri J-M. Thermodynamic promotion of carbon dioxide-clathrate hydrate formation by tetrahydrofuran, cyclopentane and their mixtures. International Journal of Greenhouse Gas Control. 2013;**17**:397-410

[62] Dashti H, Yew LZ, Lou X. Recent advances in gas hydrate-based CO_2 capture. Journal of Natural Gas Science and Engineering. 2015;**23**:195-207

[63] Partoon B, Javanmardi J. Effect of mixed thermodynamic and kinetic hydrate promoters on methane hydrate phase boundary and formation kinetics. Journal of Chemical & Engineering Data. 2013;**58**(3):501-509

[64] Mohammadi AH, Richon D. Phase equilibria of clathrate hydrates of tetrahydrofuran hydrogen sulfide and tetrahydrofuran methane. Industrial & Engineering Chemistry Research. 2009;**48**(16):7838-7841

[65] Kerenskaya G, Goldschleger IU, Apkarian VA, Fleischer E, Janda KC. Spectroscopic signatures of halogens in clathrate hydrate cages. 2. Iodine. The Journal of Physical Chemistry A. 2007;**111**(43):10969-10976

[66] Schofield DP, Jordan KD. Theoretical investigation of the electronically excited states of chlorine hydrate. The Journal of Physical Chemistry A. 2007;**111**(32):7690-7694

[67] Ramya KR, Venkatnathan A. Density functional theory study of oxygen clathrate hydrates. Indian Journal of Chem. 2013;**52**:1061-1065

[68] Nguyen AH, Molinero V. Stability and metastability of bromine clathrate polymorphs. The Journal of Physical Chemistry B. 2013;**117**(20):6330-6338

[69] Bernal-Uruchurtu MI, Janda KC, Hernandez-Lamoneda R. Motion of Br2 molecules in clathrate cages. A computational study of the dynamic effects on its spectroscopic behavior. The Journal of Physical Chemistry A. 2015;**119**(3):452-459

[70] Ochoa-Resendiz D, Batista-Romero FA, Hernandez-Lamoneda RÃ. Communication: Evidence of halogen bonds in clathrate cages. The Journal of Chemical Physics. 2016; **145**(16):161104

[71] Dureckova H, Woo TK, Udachin KA, Ripmeester JA, Alavi S. The anomalous halogen bonding interactions between chlorine and bromine with water in clathrate hydrates. Faraday Discussions. 2017;**203**:61-77

[72] Anthonsen JW. ChemInform abstract: The Raman spectra of some halogen gas hydrates. Chemischer Informationsdienst. 1975;**6**(29):175-178

[73] Janda KC, Kerenskaya G, Goldscheleger IU, Apkarian VA, Fleischer EB. UV-visible and resonance Raman spectroscopy of halogen molecules in clathrate hydrates. In: Proceedings of the 6th International Conference on Gas Hydrates. 2008

[74] Kerenskaya G, Goldschleger IU, Apkarian VA, Janda KC. Spectroscopic signatures of halogens in clathrate hydrate cages. 1. Bromine. The Journal of Physical Chemistry A. 2006; **110**(51):13792-13798

[75] Goldschleger IU, Kerenskaya G, Janda KC, Apkarian VA. Polymorphism in Br_2 clathrate hydrates. The Journal of Physical Chemistry A. 2008;**112**(5):787-789